3D Printing for Product Designers

3D Printing for Product Designers closes the gap between the rhetoric of 3D printing in manufacturing and the reality for product designers. It provides practical strategies to support the adoption and integration of 3D printing into professional practice.

3D printing has evolved over the last decade into a practical proposition for manufacturing, opening up innovative opportunities for product designers. From its foundations in rapid prototyping, additive manufacturing has developed into a range of technologies suitable for end-use products. This book shows you how to evaluate and sensitively understand people, process, and products and demonstrates how solutions for working with additive manufacturing can be developed in context. It includes a practical, step-by-step plan for product designers and CEOs aimed at supporting the successful implementation of 3D printing by stakeholders at all levels of a manufacturing facility, tailored to their stage of technology integration and business readiness. It features a wide range of real-world examples of practice illustrated in full colour, across industries such as healthcare, construction, and film, aligning with the strategic approach outlined in the book.

The book can be followed chronologically to guide you to transform your process for a company, to meet the unique needs of a specific client, or to be used as a starting point for the product design entrepreneur. Written by experienced industry professionals and academics, this is a fundamental reference for product designers, industrial designers, design engineers, CEOs, consultants, and makers.

Jennifer Loy is Professor of Digital Business Innovation at Griffith University, Australia, 3D printing specialist, and a product designer by training and at heart.

James Novak is Senior Research Fellow – CranioFacial Biofabrication at the Herston Biofabrication Institute, adjunct lecturer at The University of Queensland, Australia, and self-confessed 3D printing geek with a background in product design and architecture.

Olaf Diegel is Professor of the Creative Design and Additive Manufacturing Lab; at the University of Auckland, in New Zealand, a principal author of the Wohlers report, and an avid product designer.

3D Printing for Product Designers

Innovative Strategies Using
Additive Manufacturing

Jennifer Loy, James Novak, and Olaf Diegel

Routledge
Taylor & Francis Group

LONDON AND NEW YORK

Cover image: © James Novak

First published 2023
by Routledge
4 Park Square, Milton Park, Abingdon, Oxon OX14 4RN

and by Routledge
605 Third Avenue, New York, NY 10158

Routledge is an imprint of the Taylor & Francis Group, an informa business

British Library Cataloguing-in-Publication Data
A catalogue record for this book is available from the British Library

ISBN: 978-0-367-64111-5 (hbk)
ISBN: 978-0-367-64110-8 (pbk)
ISBN: 978-1-003-12220-3 (ebk)

DOI: 10.4324/9781003122203

Typeset in Univers
by SPi Technologies India Pvt Ltd (Straive)

For product designers, everywhere

Contents

List of figures

List of tables

Acknowledgements

The authors would like to thank the many willing (and occasionally unwilling!) participants involved in industry workshops, educational programs and research collaborations over the years that helped shape the ideas contained within this book. We must also acknowledge the inspiring, welcoming and proactive 3D printing community who make this technology as exciting as it is.

The authors would also like to thank the Business Strategy and Innovation Department at Griffith University, the Herston Biofabrication Institute and Metro North Health, and the Creative Design and Additive Manufacturing Lab at the University of Auckland. We would like to express our appreciation for the support and tolerance shown by colleagues and students (and family!) for our enthusiasm, bordering on obsession, for all things 3D printing.

Thanks to all the designers and companies who contributed stories and photographs for this book. Thank you to Leroy for copy editing the first draft of this book right until the final minutes of submission. And thank you to the publishing team for their support and advice throughout the process.

Author biographies

Jennifer Loy

Jennifer is Professor of Digital Business Innovation at Griffith University, Australia, and a product designer by training and at heart. Jen's passion about the potential of 3D printing for product designers began when the technology became accessible in 2010 and has become an obsession. She has championed designing with 3D printing, informed by a technical understanding of design for additive manufacturing (DfAM).

Jen has built expertise within several leading product design, industrial design, and engineering programs in Australia, and she has led multiple 3D printing exhibitions, conferences, and forums to foster a shared understanding of the technology and its potential. She is an invited keynote internationally on designing with 3D printing and runs workshops nationally and internationally on integrating and maximising the technology. Her research, workshops, presentations, and collaborations involve multiple stakeholders across organisations and industry sectors. These include clinical and healthcare settings, defence contexts, legal and business associations, and a wide range of commercial manufacturing companies.

Jen's work is future-focused, investigating potential future-practice and helping clients manage the disruptions that digital technologies, including 3D printing, can bring. Her aim is to support product designers and businesses through the challenges and opportunities of designing in a digital era.

James Novak

James is a self-confessed 3D printing geek whose 3D printed products have been exhibited around the world at venues including the Red Dot Design Museum (Germany) and BOZAR Centre for Fine Arts (Belgium). His research into the technology has received international awards like the Dick Aubin Distinguished Paper Award at RAPID, North America's largest annual 3D printing conference and expo. He has also been recognised by the Queensland Government for his ongoing efforts to educate the community about 3D printing and associated digital technologies, having been named an Advance Queensland Community Digital Champion in 2016.

Professionally, James has been a lecturer and researcher at 3 Australian universities, an architect, a product designer, and now a biomedical designer. He currently leads a multimillion-dollar program of research at the Herston Biofabrication Institute, an advanced manufacturing and research facility based within one of Australia's largest hospital and health systems - Metro North Health. He also holds an adjunct position at The University of Queensland and is an Assistant Editor of the Computer-Aided Design and Applications journal.

James has been using 3D printers since 2009 and actively publishing research and applications of 3D printing since 2014. He also runs a successful blog documenting many of his 3D printing projects (www.edditive.com), and freely shares many designs with the 3D printing community.

Olaf Diegel

Olaf is both an educator and a practitioner of additive manufacturing (AM, 3D printing) and product development with an excellent track record of developing innovative solutions to engineering problems. In his role as Professor of Additive Manufacturing at the University of Auckland in New Zealand, he is involved in all aspects of AM and is one of the principal authors of the annual *Wohlers Report*, considered by many to be the bible of AM. His current main area of research expertise is in design for AM.

In his consulting practice he develops a wide range of products for companies around the world. Over the past three decades he has developed over 100 commercialised new products including innovative new theatre lighting products, security and marine products, and several home health monitoring products. For this work, he has received over 50 product development awards.

Over the last 20 years, Olaf has become a passionate follower of 3D printing. He believes it is one of the technologies that has been invaluable to innovation as it allows designers and inventors to instantly test out ideas to see if they work. It also removes the traditional manufacturing constraints that have become a barrier to creativity and allows us to get real products to market without the normally high costs that can become a barrier to innovation. In 2012, Olaf started manufacturing a range of 3D printed guitars (www.oddguitars.com) that has developed into a successful little side-business.

Introduction

3D printing has evolved over the last decade into a practical proposition for manufacturing, opening innovative opportunities for product designers (also called industrial designers). From its foundations in rapid prototyping, additive manufacturing has developed into a range of technologies suitable for end-use products. Machines and software now offer a myriad of solutions to manufacturing in layers, using a wide variety of materials, many of which are unique to the technology. Furthermore, innovative individual examples of practice help illustrate the potential of 3D printing to change the nature of manufacturing. 3D printing (which will be used interchangeably with "additive manufacturing") is a technology which is part of a new digital revolution.

Whilst additive manufacturing technology is being used in some engineering and medical applications, there is a gap between the rhetoric of what is possible with 3D printing for mainstream manufacturing and the reality of the experience for professional product designers. The widespread adoption of 3D printing into mainstream manufacturing will depend on more strategic thinking about how to integrate the technology into established production practices. People, process, and product all need to be understood and evaluated from a product designer's perspective, with solutions for working with additive manufacturing developed in context. Product designers need to start with strategic business thinking within a company or consultancy, identifying at which stage of adoption a company may be, where the benefits and drawbacks may be for that company, and what is the most successful way to achieve a level of integration appropriate to their situation.

The central premise of this book is that 3D printing can be mapped onto a continuum relating to the level of adoption of the technology by a company, and the strategies product designers and businesses can implement as they build their experiences in working with the technology. This book provides product designers and business strategists with practical innovation strategies to enable their work to align with the realities of innovation and technology, as well as informing them on how they can create synergy with manufacturers and business systems to support the integration of additive manufacturing in practice. Further, the book discusses how 3D printing could contribute to society's

DOI: 10.4324/9781003122203-1

meeting the growing sustainability imperative, for example, through improving products and reducing costs in vital services such as healthcare, and for critical products within a disrupted supply chain. It further aims to close that gap with practical strategies for product designers (including industrial designers, design engineers, design consultants, and designer-makers) to support the positive adoption and integration of 3D printing into professional practice.

These strategies can be followed chronologically to transform practice for any company adopting additive manufacturing technology. Alternatively, the approaches within the three strategies described in the book can be individually selected by a product designer to meet the needs of a client. Finally, they can be used as starting points for the product design entrepreneur in building a digital business innovation strategy that maximises the opportunities disruptive digital technologies, such as 3D printing, can bring.

Overall, this book provides:

- A strategic approach to working with companies to evaluate their position in relation to the use of 3D printing in manufacturing.
- A practical, step-by-step plan for product designers aimed at supporting the successful adoption of 3D printing by stakeholders in all levels of a manufacturing facility, tailored to their stage of technology integration and business readiness.
- An analysis of real-world examples of practice in relation to the strategic approach outlined in the book.
- An enduring resource that can be used over the coming years to transform practice, and that does not rely on specific hardware, software, or materials.

The book is written for product designers by product designers, and it evaluates the opportunities and challenges of additive manufacturing from that perspective.

CURRENT CONTEXT

The global economy is being transformed through the paradigm shift that digital convergence is creating. Research on workforce development for a digital economy needs to focus more on the impact digital technology should be having on content and educational intent. Nigel Cameron, in his book on the rise of automation, suggests that there is insufficient preparation for the digital future, stating:

> Our leaders need to be able to think on new, fast-moving timetables. The education/skills training implications of engaging seriously with the disruption scenario we are discussing stand out as immediate challenges for policymakers thinking ahead. The Industrial economies have never experienced anything like it.

[1]

According to climate change activist Al Gore [2], "technology-driven changes are now playing a much larger role in determining the future of work." His view is that:

> National policies, regional strategies, and long accepted economic theories are now irrelevant to the new realities of our new hyper-connected, tightly integrated, highly interactive, and technologically revolutionized economy. … the global economy is being transformed by changes far greater in speed and scale than any in human history.

Cameron supports this view, raising further concerns over the preparations that need to be made in anticipation of changes for the future workforce: "On any reading of the situation, our leaders need to prepare for a period of perhaps unprecedented turbulence in labor markets. The evidence is that they aren't."

Additive manufacturing for product designers, as opposed to engineers, should not be seen as an individual set of fabrication technologies, but rather an integral part of the digital convergence of connected devices, data generation, artificial intelligence, machine learning, and digital fabrication. This places it firmly in the middle of the technology-driven changes that Gore described. Learning to work with 3D printing will, therefore, be effective only if that learning extends beyond the mechanical requirements of achieving a print. It needs to encompass the implications of working within a changed, digital landscape. Understanding design for 3D printing, in this context, involves understanding the impact that digital convergence is having on the role of workers on the factory floor. It involves being sensitive to the disruption and insecurities that radical changes in established practices can bring, and having a realistic understanding of the financial commitment associated with introducing industrial additive manufacturing into a traditional production environment.

For the product designer, the development of a successful product using industrial-level additive manufacturing for a company involves more than getting the product details right. It is more than having sound knowledge of the specific additive manufacturing technology being used, or knowledge of the materials. It is more than knowing the workflow, file preparation, or how to optimise the components, despite the importance of these factors. It is also about making sure the product being proposed is right for that company, at that time, based on a critical evaluation of the company's in-house expertise, its willingness to invest in the technology and peripheral systems, its supply chain, and its adaptability to additive manufacturing materials and distribution.

Business strategist Peter Drucker [3] emphasised the importance of "doing the right thing" before focussing on "doing it right." In design for additive manufacturing, it is critical to be creating the right product for a company's particular stage of technology adoption, prior to focusing on the details of effective design for fabrication. Understanding what makes the product right for the company needs to be based on a review of all stakeholders. Additive manufacturing is a disruptive digital technology, and not all workers will be accustomed to the change in practice digital technology can create. Presenting a product proposal that demonstrates a good use of the technology needs to be supported with a rationale for how it will fit into an existing portfolio of practice that underpins a company's success.

> According to annual surveys by McKinsey, 70 percent of all change initiatives fail. The main barriers to successful change, nearly 60 percent, have been employees' attitudes and non-supportive behaviour by management. The situation has not improved over the years: People don't want change. They don't ask for business models which oblige them to unlearn what they have learnt. They are afraid of losing something.
>
> [4]

Nevertheless, additive manufacturing for end-use products is still an emerging technology for product designers in most fields. It is also new to most companies. Exploring the type of digital business innovation that would maximise the opportunities additive manufacturing provides in a digitally connected era would involve a company moving beyond practical adaptation to transformational practice. Working with a company on this journey of disruptive technology adoption requires a product designer to collaborate in the company's development as much as in the design of products themselves. More broadly, rethinking product design for a more sustainable future involves maximising materials and changing the relationships between people and products. As suggested by Aspelund, "Unless we begin to seriously redesign our products and methods, there will be less available for future generations" [5].

This is more pertinent in the context of recent global events, such as the COVID-19 pandemic, which have forced designers, manufacturers, and indeed nations to reassess and implement new arrangements for operating in uncertain and rapidly changing environments. Additive manufacturing played an important role in the emergency supply of medical devices and personal protective equipment, among other products, and exhibited flexibility in contrast to more established and conventional manufacturing methods [6]. Designers connected and collaborated online like never before, sharing 3D models, feedback, documentation, and other resources. As the world recovers and plans for a future in which similar global crises are possible, product designers must play a key role in framing a manufacturing industry that is responsive, flexible, and more in tune with societal needs at any point in time.

Realistically, the growth of additive manufacturing as a central manufacturing technology will face significant challenges. Development of machinery and materials tends to be proprietorial, and there is a lack of cross-pollination of operations. Some of the largest companies involved in additive manufacturing can be regularly seen buying out start-ups and smaller companies to own relevant, or even competitive, intellectual property. Knowledge sharing in an emerging industry sector is understandably limited by competition. It is also unlikely that an additive manufacturing-based economy, as posited in the sustainability discussion in Chapter 8, will happen without the influence of external forces such as a climate change crisis or additional health crisis that reconfigures the structures of societies. The development of policies such as the OECD Extended Producer Responsibility, which places the responsibility of dealing with the end of life of a product back on the manufacturer, will also impact manufacturing in a way that could bring additive manufacturing more to the fore. Additive manufacturing can enable digital entrepreneurship, and product designers will be better positioned to make and market their own products through the facilities that digital connectivity and fabrication provide. Even so, the hype around 3D printing as a "push-button" technology that requires little or no skill, and the complexity of which is "free" without any cost to the producer, sets up a false narrative that the product designer will have to overturn.

The current lessons about additive manufacturing have been dominated by the body of knowledge built within the mechanical engineering and biomedical disciplines. These fields have made considerable investment in the development of additive knowledge specific to the needs of their profession. Architecture has gradually made inroads into working with the technology in its own disciplinary interpretation. Product design, however, has been slow to commit to the technology. Desktop 3D printers are ubiquitous in product design studios and education, but beyond their role as a prototyping technology, the use of additive manufacturing for end-use products is still very much in its infancy. Product designers need a comprehensive understanding of working with additive manufacturing, informed by technical detailing but also by an understanding of the implications of working within a connected, digital manufacturing future. In *The Future Starts Here*, in relation to digital technologies, the authors state: "We can sit back and let that happen, or we can take an active role in steering where these things take us" [7]. Moving beyond the body of knowledge built within engineering and architecture on 3D printing would provide product designers with the foundations for driving design for additive manufacturing, rather than adopting practice.

SUMMARY OF CHAPTERS

Chapter 1 – Demystifying 3D printing processes and workflow

This chapter introduces the realities of working with 3D printing for product designers, specifically from their point of view. Drawing on the authors' hands-on experience with 3D printing (aka additive manufacturing) over a cumulative 40+ years, this chapter untangles the confusion between the different additive technologies, re-framing them specifically for product designers with recommendations on how to approach learning about them and integrating them into practice in a structured way. The approach described in Chapter 1 challenges conventional learning on the topic, offering new recommendations on which technologies to start working with and how to progress this learning, from those with fewest constraints and wide end-use possibilities to those with many design rules that can be complex systems to manage. It includes practical advice on differentiating between part samples from different processes, and it introduces how to approach their evaluation, including features that are good examples of design for process. It also details technical qualities of 3D printing file types (such as STL, OBJ, VRML, 3MF, and AMF), as well as the digital work-flows to go from CAD to 3D print, and strategies to repair and slice files appropriately.

Chapter 1 outlines hands-on experience in working with high-end selective laser sintering (SLS) to help to demystify this process, and it includes the key steps from preheating a SLS machine through to the final post-processing of printed parts. Behind-the-scenes photos not commonly shown in glossy sales brochures or publications help explain these processes. Lastly, the chapter pro-vides a case study of metal 3D printing (selective laser melting, SLM) of a bespoke product, including details of the labour-intensive and manual processes after a metal part is printed, and examples of pushing the boundaries with this technology. These help to explain the realities of working with high-end polymer and metal printing that are likely to be encountered in manufacturing environ-ments, with details often invisible when relying on service bureaus, and they provide a basis for designing for process in the later chapters.

Chapter 2 – Working with a design for additive manufacturing (DfAM) consultancy

This chapter introduces the three strategies that form the foundation of this book:

- Strategy 1: Working with existing production.
- Strategy 2: Product redesign and new product design.
- Strategy 3: Digital business innovation.

In addition to developing design *for* additive manufacturing skills sufficient for designing *with* 3D printing, the authors recommend that product designers work on strategic business thinking within a company. This involves identifying at which stage of adoption of 3D printing a company may be, where the benefits and drawbacks may be for that company in investing in the technology, and what would be the most successful way to achieve a level of integration appro-priate to their situation.

The second part of the chapter then describes the conversations of an imagined manufacturing company working through the three stages of 3D print-ing technology adoption with the advice of product designers from a fictional design consultancy. The intent is to frame the pathway a company might take, along with articulating some of the problems the product designer might face as a consultant or in-house designer, such as resistance to change from within the company, and a lack of understanding of the complexities involved in integrating the technology into a business model. The script of these conversations follows key stakeholders in the organisation as they meet over time with their consultant to discuss the stage of technology adoption that the company has reached. The fictional firm is used throughout later chapters to help illustrate how a product designer could implement some of the strategies with companies of different size and motivation.

Chapter 3 – Strategy 1: Working with existing production

Chapter 3 introduces the initial stages of a progressive, cumulative strategy to support the enthusiastic adoption of 3D printing into an established traditional manufacturing facility – not an easy task. Divided into six stages, the approaches discussed at each level of this first strategy are designed to engage and influence the attitudes of a manufacturing workforce that is not familiar with 3D printing and potentially resistant to adoption. The intent is to identify concerns of the workforce in relation to additive manufacturing and evaluate the level of expertise, in preparation for the introduction of non-threatening, positive uses for the technology integrated with their existing production methods.

The strategy starts with the most easily accepted approach to using 3D printing in an existing manufacturing practice, building from visual prototyping, test prints, and iterative design to the use of the technology for bridge manufacturing, testing the market prior to conventional mass production of a product. From that base point, the recommendation is to introduce 3D printed jigs and fixtures to production as a way of using the technology to support existing practice, building workforce familiarity. It also results in personal benefit for the workforce – for example, improved safety and ergonomics – and demonstrates the use of personalisation through additive manufacturing, potentially opening new markets. Once the workforce is enthusiastic about using the technology in this context, the next step of Strategy 1 is to enhance tooling using additive manufacturing for added value – for example, adding conformal cooling to a tool, providing quantifiable improvements to production. The last step in this strategy is to introduce agile or flexible tooling using 3D printing inserts, or short-run, stand-alone 3D printed tooling. Depending on the manufacturer, product designers may incrementally lead the adoption of these approaches, or implement one approach that presents the best business case. The goal is to initiate adoption with minimal disruption before advancing to the more disruptive approaches detailed in Strategy 2 (Chapter 4).

Chapter 4 – Strategy 2: Product redesign and new product design

Once the workforce and executives of a manufacturing company have a positive, open attitude towards 3D printing, after having explored the first strategy outlined in this book (Chapter 3), the professional product designer can lead product redesign and the design of completely new products appropriate for additive manufacturing. This is the collection of approaches that often leads to the eye-catching 3D printed products shared in the media, and the one many designers and engineers want to jump to. However, skipping over the first strategy to this more challenging second strategy is a mistake which may cause resistance from an established workforce, as well as problems gaining support for the inclusion of the technology on the shopfloor. It requires consideration of a company's technical capability, as well as new equipment, logistics, and customer demand.

Once a company is ready to embrace 3D printing for end-use products, the approaches included in Strategy 2 progressively advance from redesigning existing products through to completely new products and systems to support them. This begins with the technical optimisation of existing products through light-weighting, topology optimisation, and part consolidation. This includes opportunities to apply generative design workflows, finite element analysis, and lattice structures through advanced design software. These approaches build through the adoption of customisation strategies that may allow for some level of customer interaction and input, towards the most radical and potentially impactful applications of new product innovation and forms driven by functional requirements. Examples of these approaches are illustrated through the work of leading product designers specialising in new product innovation using 3D printing.

Chapter 5 – Strategy 3: Digital business innovation

The final strategy of this book goes beyond 3D printing of new or reimagined products detailed in Strategy 2 (Chapter 4) and begins to work through issues around the relationship between the customer and the producer, its potential for change, and the operational impacts of introducing 3D printing into an existing business. Through the five approaches detailed in this chapter, the often-unanticipated consequences and disruptive changes caused by shifting to additive manufacturing are discussed.

Strategy 3 begins with digital inventory and new methods of distribution enabled by 3D printing, pertinent considerations following the COVID-19 pandemic and global disruption to manufacturing and supply chains. Personalisation and scalable systems of supply then build on these supply considerations and go beyond the confines of customisation, integrating with personalised data such as 3D scans and medical images, whilst accommodating for "markets of one." Finally, digital business innovation represents the culmination of all previous approaches through "Strategy 1: Working with existing production," "Strategy 2: Product redesign and new product design," and Strategy 3. Numerous examples from industry are provided to illustrate how entire industries have been transformed by considered and structured adoption of 3D printing. This chapter is also important for entrepreneurs and start-ups whose agility allows them to begin working at the cutting edge of additive manufacturing unencumbered by traditional manufacturing paradigms and may be an appropriate starting point informed by the approaches of previous strategies.

Chapter 6 – Case studies: 3D printing from the product designers'
perspective

Chapter 6 provides over 25 diverse examples of product design practice, demon-strating how 3D printing has been implemented using one or more of the approaches defined within the three strategies of the previous chapters. The focus is on inspirational and unusual avenues for working with additive manufacturing for the product designer and includes high-quality photographs and behind-the-scenes details of the realities of working with the technology. Some of the examples are from the authors' own practice, such as the world's first full-size 3D printed bicycle frame in one piece (James Novak), and a decade of 3D printed guitars (Olaf Diegel), while others are from high-profile companies and international designers like 3D printer manufacturer Renishaw, and Lionel Dean from FutureFactories. The intent is to encourage the exploration of new applications of the technology and adven-tures in design and practice. The eight case study topics are:

1. Tougher and faster: Additive manufacturing for bicycles.
2. Additive manufacturing of musical instruments: Guitars.
3. 3D printing for the designer-maker: furniture.
4. New ways of working: Film industry – animation, props, and costumes.
5. 3D printing personalised design for health and well-being.
6. Additive manufacturing medical devices.
7. Creative exploration: Playing at the boundaries.
8. Customisable technical products.

Chapter 7 – DfAM: Design guidelines for product designers

Chapter 7 provides product designers with clear guidelines on design for addi-tive manufacturing (DfAM), written with their workflow in mind. It includes ref-erence guides and summary tables of key features and specifications for the most relevant additive manufacturing technologies to product designers, expanding on the introductory information in Chapter 1. This includes compari-sons of build volumes, recommendations for minimum wall thickness, toler-ances for 3D printed assemblies, designing for colour 3D printing processes, and other important technical guidelines.

In addition, Chapter 7 includes practical information on pre-production preparation, including optimisation and file preparation. It also discusses design-ing for post processing, which is an area often overlooked in instruction manuals and news articles promoting the latest 3D printing success stories, but critical for product designers and companies adopting the technology. This includes designing to minimise post processing and a summary of the post-processing treatments that can be used to enhance the performance characteristics of parts after 3D printing. This ranges from paints and dyes to vapour smoothing, bead blasting, polishing, metal plating, and heat treatments. Photographs include both good and bad examples of DfAM, as well as parts before and after post-processing and cross-section views of 3D prints to show internal details.

Chapter 8 – 3D printing sustainability and digital ecosystems

Environmental, economic, and social sustainability are an increasingly global issue, with environmental, health, and political pressures impacting the way manufactured goods are made and distributed. Chapter 8 considers the potential of 3D printing to contribute to future proofing manufacturing in the face of short-term, emergency production demands. It discusses paradigm changes to production systems to help sustain manufacturing when material supply and downstream demand are both unstable and designing for the circular economy and extended producer responsibility. The chapter also provides an overview of how global megatrends have influenced the impact of 3D printing for sustainability and could potentially impact the role and output of designers in the future. In addition, the chapter highlights the societal imperative for revising existing production systems and attitudes. It draws a line under the industrial revolution and current mass production, discussing reasons for a radical rethink of design, supply, consumption, and product disposal based on additive manufacturing.

Chapter 9 – Making the future/remaking product design

3D printing technology is now over 30 years old. Chapter 9 is a call to arms for product designers to lead the development and implementation of 3D printing in a world facing the new and fluctuating challenges described in Chapter 8 as well as throughout the book. It summarises the key barriers for 3D printing adoption that need to be overcome and recommends that product designers construct effective frameworks to maximise additive manufacturing for manufacturing clients. Alongside the advantages to individual companies, this approach has the potential to provide global societal benefits, as consumerism shifts beyond mass production to a personalised, print-on-demand, value-added system. The cost to the environment of established mass-production practice is too high, and a more value-added, print-on-demand approach will help to redirect the economy. Yet these ideals contrast with a current adoption of 3D printing, driven by equipment manufacturers. Beyond engaging with this book, what can product designers do to transform manufacturing for the twenty-first century? How will we look back on this time in another 30 years of 3D printing technology developments? This chapter spurs broader thinking about what is an exciting technology, but ultimately a technology that is directed by people and companies. Can we do better? The chapter concludes a book that is designed to provide product designers with the tools and structure to lead the appropriate adoption of 3D printing technology, working realistically with the manufacturing systems in place whilst steering the stakeholders, from machinists to customers, towards a more sustainable and enabled future.

REFERENCES

1. Cameron N. *Will Robots Take Your Job?* Cambridge: Polity; 2017.
2. Gore A. *The Future*. New York: WH Allen; 2013.

3. Drucker P. *The Effective Executive: The Definitive Guide to Getting the Right Things Done*, revised ed. New York: HarperBusiness; 2006.

4. Gassmann O, Frankenberger K, Csik M. *The Business Model Navigator*. London: Pearson; 2014.

5. Aspelund K. *The Design Process*, 2nd ed. London: Fairchild Books; 2010.

6. Novak JI, Loy J. A Critical Review of Initial 3D Printed Products Responding to COVID-19 Health and Supply Chain Challenges. *Emerald Open Research* 2020;2(24). doi:10.35241/emeraldopenres.13696.1.

7. Hyde R, Pestana M. *The Future Starts Here*. London: V&A Publishing; 2018.

1 Demystifying 3D printing processes and workflow

- From "design for manufacturing" to "design for additive manufacturing."
- Cutting through the confusion: A product designer's view of additive manufacturing technology.
- At a glance: Visual identification starting points for recognising processes from parts.
- Design rules for software – workflow and file preparation.
- Getting hands-on with high-end industrial polymer printing: Key stages of the workflow designers need to know.
- An introduction to metal 3D printing for product designers, including the realities of post-processing.

MOVING FROM "DESIGN FOR MANUFACTURING" TO "DESIGN FOR ADDITIVE MANUFACTURING"

How can product designers evaluate the differences between a product that is well designed and one that is not? What judgements can be applied? Arguably, a good design addresses the functional and emotional requirements of its purpose for the target market and responds to aspirations for society at a particular point in time. However, it needs to be designed properly for the manufacturing processes to be appropriate to the industrial context. Yet, over the last century, product designers have been trained to design products for mass-production technologies. This is evidenced by the fact that one of the most widely used methods by professional designers to evaluate the knowledge and experience of a graduate is to ask for a design concept specific to a particular production process. It is possible to then judge the appropriateness of that design for the constraints and opportunities of that manufacturing process. As product designers know, for example, there are characteristics to design for injection moulding. At its most basic, these characteristics include the design of draft angles, to allow the part to be removed easily from the mould, and parting lines, to demonstrate an understanding of the pathways of a mould as it is pulled apart. Therefore, avoiding details on undercuts, for example, is an indication of knowledge of the process.

DOI: 10.4324/9781003122203-2 **12** □

The level of sophistication in the design beyond these fundamentals is an indication of how well the designer understands the process and responds to it. Designing to accommodate parting lines, and the removal of flashing, suggests experience with the technology. The appropriate specification of injector pin placement can indicate a further understanding of the design of tooling for the process.

Taking that expertise to the next level, understanding multi-part moulds, the costs and difficulties that can be involved, and the need to work with the manufacturer to develop the product and tooling, rather than develop a concept in isolation from the manufacturing process, is integral to professional product design practice. Beyond this, however, product design expertise in designing for injection moulding must not only meet the practical requirements of specification for the process but also maximise the technology to create innovative products that exploit the technology's particular characteristics. Production moulds can be extremely expensive, and the product designer needs to be aware of the cost implications of a design where moulds need to be fabricated. If a new mould is to be created, the product designer needs a good justification for that investment by the company.

Digital fabrication is different. It started in mass production in the middle of the twentieth century as computer numerically controlled (CNC) routing. As with injection moulding, designing for CNC routing requires understanding of the technology. At a fundamental level, this is where the viability of proposed parts and products can be judged on whether they can be cut using a CNC router in 2D or a CNC router of more than three axes, and whether the designer understands the differences. As with injection moulding, the pathway of the cutter is a critical element in the design. In addition, accounting in the design for the width of the cutter and the position of the cutter on the tool path indicates the product designer's experience with the technology.

As with designing for injection moulding, designing for CNC routing, beyond the basics, involves demonstrating the ability to exploit the technology's characteristics to maximise its use. For 2D digital fabrication using CNC routing, for example, this would also involve understanding nesting parts on a board to maximise the material use, the gaps needed between parts to maintain the board integrity for removal, and the size of the board used by the company – usually a full-sheet 2400 × 1200 mm, but can be a half sheet or smaller depending on the machine. Added sophistication comes from the specification of the tabs that hold parts in place during cutting. These half-depth cuts (this can vary depending on the size of the parts and thickness of the board) can be automatically generated, but they may then appear on the curve of a part, or the front face. The straight edge of the tabs may cause additional cost in post-processing, if visible, or potentially lead to quality control defects where tabs must be smoothed off curved edges.

However, the most significant difference in digital fabrication as opposed to conventional mass-production techniques is that it allows for unique parts to be cut as there is no initial investment in tooling prior to part manufacturing, which has been an influence in terms of what is designed for CNC. This helps overcome some of the great barriers to getting a product to market (and to innovation), the lead time, and cost of making traditional production tools. In some cases, additive manufacturing may be able to help reduce this barrier and significantly reduce lead times.

Additive manufacturing is a development of computer-driven fabrication. While it was developed in the mechanical engineering discipline, predominantly for the aerospace and automotive industries and for medical applications, over the last decade it has become prevalent in product design. It was first introduced as a prototyping technology for creating proof-of-concept models for products to be made using traditional manufacturing technologies. At that point, they were not designed specifically for additive manufacturing processes, as they were mimicking traditional processes. Over the last decade, however, significant developments in additive manufacturing technology have taken it from a predominantly prototyping and visualising technology to one suitable for functional parts. This raises its importance for the product design profession.

Thus, for product designers, designing for additive manufacturing now needs to be approached from a product disciplinary perspective and not as a by-product of engineering research. Whilst there are many resources written by engineers, for engineers (listed as resources at the end of this book), designing for additive manufacturing for product designers is different. There are differences in priorities, differences in approaches to the technology, and differences in the technology's integration into practice. As with traditional production technologies, good design for additive manufacturing (DfAM) can be judged at a fundamental level, where the part is possible to be made using an additive manufacturing technology. It can also be judged at a more sophisticated level, where the products maximise the opportunities of the process whilst working within its constraints.

As product designers, the introduction of the technology into education has, arguably, been at odds with the way it should be approached in practice. Where desktop single-filament 3D printing tends to be the first hands-on experience of the technology for product designers, it has caused some difficulties in its adoption in professional practice. This is because the design rules for working with single-filament printing and other technologies, such as polymer powder-based laser sintering technologies, are very different. For example, the nature of single-filament 3D printing means that designing for this, as opposed to simply prototyping, can be more restrictive and requires greater knowledge than designing for powder-based laser sintering. For a professional or aspiring product designer, or for a product design academic engaged in professional development, the introduction of additive manufacturing through single-filament printing is at odds with how DfAM should logically be learned and has largely been driven by the low-cost and wide-spread access to this technology.

The approaches to integrating the technology into industrial practices described in this book are based on product design practice principles and work with disciplinary practice. Similarly, in discussing the technologies themselves, this chapter approaches describing the processes and introducing professional product designers and students to their constraints and opportunities, through the lens of product design practice, not engineering. Once this bridge has been established in industry, engineering resources can be more effectively adapted for product design practice.

The following chapters introduce a technology adoption pathway developed with a product designer's disciplinary knowledge and approach in mind. The pathway provides the product designer with an approach that works from the most straightforward process to design for, through to the most technically difficult. It works with product design conventions rather than engineering ones, or those that have emerged from the maker-space movement. For product designers, understanding the technology covered by the terms "additive manufacturing" and "3D printing" in the current context needs a discipline-specific approach to understanding, and working with, the individual processes. These sections provide product designers with that introduction. They also enable designers to recognise the characteristics of parts produced using different industry-level additive manufacturing technologies, and a framework for evaluating good design outputs in each case. These sections are followed by two practical examples of working hands-on with the technology for product designers to help familiarise designers with what to expect when faced with high-end, industry-level, additive manufacturing technologies in-house.

CUTTING THROUGH THE CONFUSION: A PRODUCT DESIGNER'S VIEW OF ADDITIVE MANUFACTURING TECHNOLOGY

There are currently seven broad categories of additive manufacturing technologies, defined by ISO/ASTM 52900 standards (version 2021 at the time of writing) [1]. The order of the process categories presented in the international standards has not been developed with design or learning in mind, simply ordered alphabetically. Therefore, it does not provide an indication of which are the most straightforward to work with, and which are the most complex. In this section, the ISO/ASTM 52900 standards are outlined for reference, and then a revised approach to the technologies is explained to better support the professional development of designers and product design academics.

1. Binder Jetting.
2. Directed Energy Deposition.
3. Material Extrusion.
4. Material Jetting.
5. Powder Bed Fusion.
6. Sheet Lamination.
7. Vat Photopolymerisation.

ISO is the International Organization for Standardization, a familiar governing body for product designers who provide specifications and guidelines for materials, products, processes, and services. ASTM International, formerly known as the American Society for Testing and Materials, is a similar international organisation responsible for a broad range of standards, initially formed by a group of scientists and engineers in 1898.

ISO/ASTM 52900 aims to standardise terminology and concepts associated with additive manufacturing, ensuring clear communication among people involved in the technology. Most books published on additive manufacturing expand on the technologies using these same categories. However, by reducing varied and complex processes into just seven categories, vital information for product designers is lost and can be difficult to unpack when learning about additive manufacturing, particularly when learning may be reliant on available technologies in a classroom, rather than based on levels of complexity or applicability to end-use manufacturing.

The categories are also defined by the mechanical process through which they additively produce a part, which does not specifically relate to materials, design restrictions, or other information that may be important when a designer is creating a new part for an additive process. For example, the most common process, material extrusion, can relate to materials as varied as metals, polymers, foods, human cell solutions, and concrete, yet it would be unusual for designers in industry to have genuine expertise in such a range of materials if they worked within traditional manufacturing settings. Knowledge in extruding plastic may have similarities to extruding solutions of human cells. However, it is unlikely that a designer used to working with plastics would one day walk into a laboratory and be able to successfully 3D print solutions containing human cells without significant training. Furthermore, with the rapid pace of development of additive technologies, there are an increasing variety of novel processes that do not fit into any of the seven categories yet may be of interest to designers. Therefore, it is important to consider a different model of defining additive technologies, one that considers the relationships between manufacturing processes and materials and provides guidance about the complexity of designing parts for production. Figure 1.1 displays a new framework for product designers to understand additive technologies which will be unpacked throughout this chapter.

Necessary generalisations are included in the formulation of Figure 1.1, and the determination of the order in which AM technologies, and groups of technologies, are arranged. Nevertheless, the diagram provides a high-level perspective of the complexity specifically for the product designer, from least to most complex, when read from top to bottom and from left to right. For example, as a rule, non-metals are easier to design for than metals. Within each category this principle also holds true, for example, in the top row, selective laser sintering is easier to design for than multi jet fusion. The separation of designing for metals compared with other materials is an important distinction that is critical when

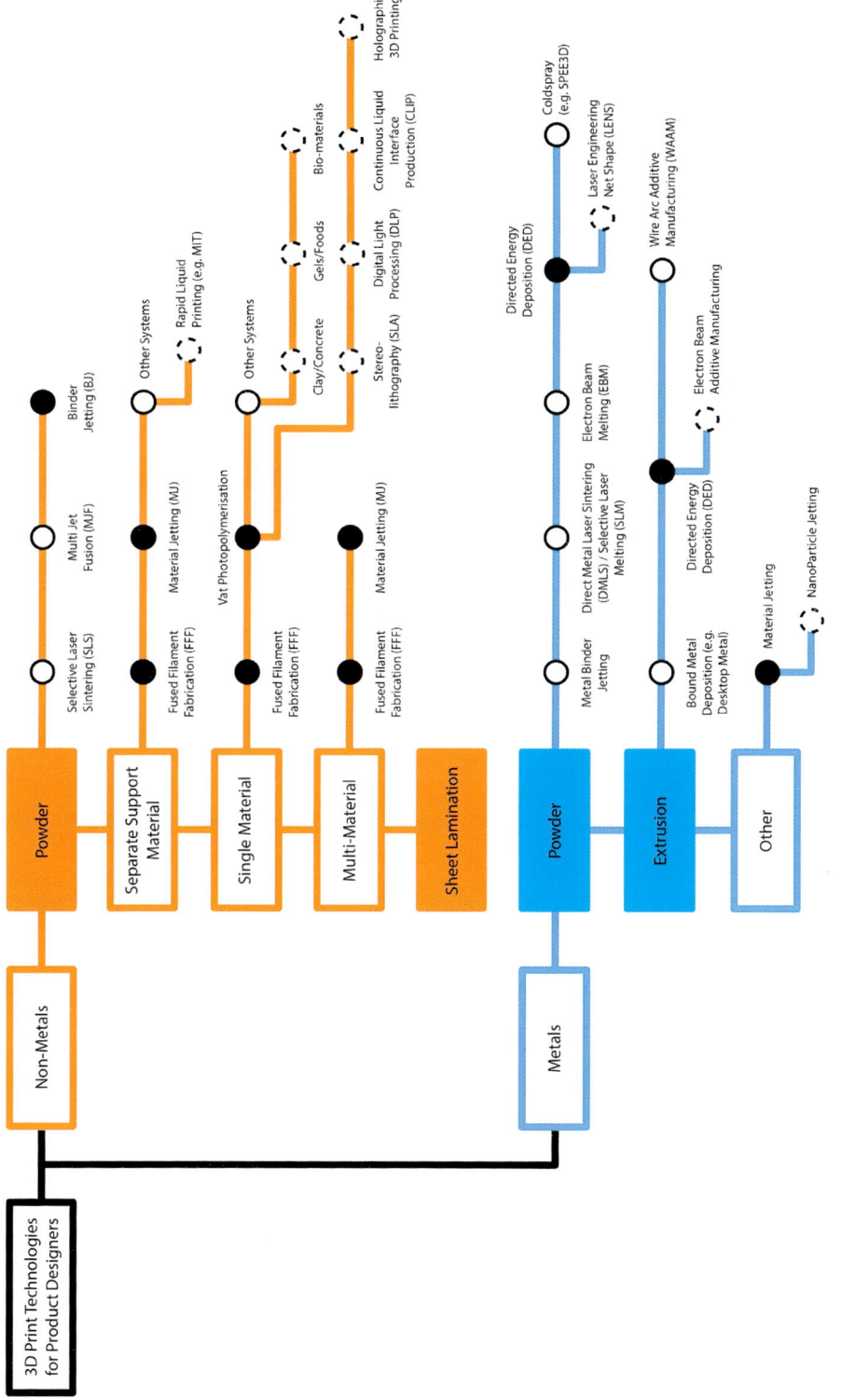

1.1

A new framework for additive manufacturing technologies for product designers. Solid boxes/circles indicate overlap with the ISO/ASTM 52900 standard categories of additive technologies.

reading Figure 1.1. It recognises the complexity of all stages of designing, pre-processing, printing, post-processing (including heat treatment and potential for shrinkage), and validating metal components. In simple terms, this means that a designer with no experience using AM technology would more readily learn how to design a part for polymer technologies, like selective laser sintering, compared to what can appear to be the very similar metal technology of selective laser melting. Within ISO/ASTM 52900 standards, both are part of the same category.

Whilst this approach is intended to be indicative of the ease in which a designer may learn a specific AM technology through design, it is certainly not necessary to learn 3D printing beginning with SLS and proceeding sequentially down the technologies outlined in Figure 1.1. For example, it is possible to learn metal 3D printing in lieu of other technologies, especially in situations where a designer works in industries like aviation or automotive where there may be an exclusive focus on metal additive manufacturing. In this case, investing the time and resources to learn polymer processes may not be viable. However, it is suggested that a designer learns the metal powder processes in preference to metal extrusion processes due to the design freedoms of powder processes as well as the quality of surface finishes compared to extrusion processes, which typically require significant post-processing. Yet, a product designer looking to gain broader expertise in AM, or a design student preparing for the future workforce, should begin by learning polymer processes like SLS and MJF, as these offer design freedoms for a good range of commercial applications. Mastering these processes prior to some of the more complex ones, which require additional design constraints, will build the designer's knowledge and confidence before working with additive manufacturing processes that are suitable to more niche design fields.

Figure 1.1 also highlights which processes or categories are directly related to the ISO/ASTM 52900 standards. As a result of separating metal and non-metal processes, technologies like powder bed fusion and material extrusion appear several times, no longer overarching categories but distinct in the materials, as well as the design knowledge, required to competently manufacture parts for them. "Powder" has also come to represent two previously distinct ISO/ASTM 52900 processes within the non-metals section – binder jetting has more in common with powder bed fusion technologies such as selective laser sintering and multi jet fusion when viewed from a designer's perspective. While mechanically their operation is different, and therefore separated within the standards, for designers the commonalities between them mean that knowledge translation is easy, and there is more in common between them than between polymer and metal powder bed fusion techniques. Therefore, a part designed for binder jetting could be produced using SLS with little or no change to the part, whereas a part designed for SLS may require significant modification to be successful using SLM, despite being part of the same powder bed fusion process within ISO/ASTM 52900.

Some limitations also must be mentioned in relation to Figure 1.1. First, it is important to clarify that while designing for SLS is the easiest of all AM technologies (in our view) and is identified here as a recommended starting point for the professional designer, it is certainly not the cheapest or most accessible technology. Most product designers and students will be familiar with fused filament fabrication (FFF, also known as FDM) technology, which has become ubiquitous within design studios, maker spaces, and workshops predominantly as a means of prototyping.

FFF technology is relatively simple, built from relatively cheap hardware components and often operated with free slicing software, and desktop machines are available for just a few hundred dollars, with more professional systems available for just a few thousand dollars. Therefore, a single material, or even a dual or multi-material extrusion system, may be far more accessible for product designers and may be easy to learn in a hands-on fashion. However, from a professional perspective, and a view towards end-use parts rather than prototyping, SLS offers numerous advantages. These range from the freedom to produce parts without the need for support material, to the high-quality surface finish and durable materials, which will be detailed in the following section. The printed parts look and feel like final products and, in many industries, are used as functional products. Therefore, SLS provides a valuable introduction to additive manufacturing for product design professionals and is recommended as the starting point for engaging with AM. Whilst low-cost, desktop FFF can be useful for developing test prints for SLS product development, as it is accessible and inexpensive, there are real challenges with single-filament desktop printing for professional design practice.

DESIGNER'S COMMENT

To achieve a good-quality outcome with FFF requires building a considerable body of knowledge and experience in working with the process. For example, with filament, extrusion-made, vertically printed screw posts can often see the self-tapping screw strip out the inside wall of the screw post. An effective solution to remedy this weakness is to let a few drops of super glue wick in between the contour and infill material to strengthen that feature of a part.

Working with SLS as a starting point is likely to produce a more reliable outcome, both in terms of the design being possible to make and avoiding the practical pitfalls of working with FFF. However, access to SLS as a starting process within a company – or as a design consultant or entrepreneur – may not be straightforward. Strategies a product designer may implement to produce SLS test pieces and functional parts if such technology is not readily available include:

1. *Using online or local service bureaus.* Most will provide an instant quote upon uploading a part and can ship it directly to the designer or their client. Quotes for 3D printing are typically based on the amount of material used and the overall size of the part. Therefore, if the design is clever and the volume is kept low, it is possible to reduce the cost of parts. This can be used as a learning tool even without getting parts printed. An example would be to examine how hollowing a solid section of a design affects the cost of production, compared to it being solid or using a lattice structure.

2. *Connecting with a local university or makerspace that has SLS or other new technologies in-house.* Universities often run workshops for industry and are increasingly open to responding to the needs of local business. Makerspaces are also open to the community and can provide access to a broad range of equipment, including 3D printers, although few provide 3D printing technologies beyond FFF.

3. *Leveraging contacts.* A professional designer may not be surprised to find that a supplier, client, or other contact has access to a commercial 3D printing technology, such as SLS, and is willing to provide discounted printing rates and even some hands-on training. Many companies are willing to support professional development for designers they have worked with previously. In addition, materials for many 3D printers have an expiry date (for example, material jetting), or small window in which they are optimal. So, owners of such machines are looking for extra parts to print to fill a build volume, or to print overnight to make use of the material, and the machine.

For those designers familiar with using low-cost, desktop FFF for test pieces and early prototyping, it is useful to compare parts produced with a single-filament FFF with those produced with SLS. There will be a significant difference in surface finish in different areas of geometry, fine detail (e.g., embossed text), and evidence of print orientation. The problems of removing a 3D printed support structure from around the part, which is time-consuming and can damage surface quality in FFF, are avoided completely with SLS. It will quickly become evident that SLS produces superior parts for most product applications and can cope with more complex geometries.

Whilst even technologies like SLS have their constraints and challenges, by starting with the most straightforward from a product design point of view, then working through the technologies with increasingly complex requirements, the product designer can incrementally build a body of knowledge on the technology, much as they would for conventional manufacturing technologies.

AT A GLANCE: VISUAL IDENTIFICATION STARTING POINTS FOR RECOGNISING PROCESSES FROM PARTS

Based on Figure 1.1, it is clear there are numerous and varied technologies referred to as additive manufacturing. Most designers' first encounters with the technology tend to be with a single-filament desktop FFF. However, learning how to work with the technology as a professional product designer should start with the technology with the fewest constraints for designers, then work through to the technology that has the most limiting constraints. In this way, the designer builds knowledge incrementally, much as is usual practice with conventional technologies, rather than starting with the technologies where it is the most difficult to achieve a professional output.

One of the initial, and unnerving, challenges for a product designer asked to work with additive manufacturing for a particular company for the first time is when the designer is given several different samples and asked to comment on them. It is, therefore, useful to have a way of decoding these parts and making a basic evaluation of their suitability as designs, particularly where they could relate to the practice of that company. As explained in more detail in Chapter 7, it is not the case that "anything" can be printed, or should be printed, and certainly what might be suitable to be printed through a particular 3D printing technology might equally be unsuitable for another. As a product designer, working with the constraints and opportunities of different processes is as important for additive manufacturing as it is for conventional manufacturing processes. Chapter 7 provides reference tables that describe three increasingly sophisticated levels of design for process with the aim of developing a product design approach to evaluating the expertise demonstrated in 3D printed designs.

When faced with several parts or products that are 3D printed, there are basic features that can help the product designer quickly decipher which processes have been used. This section summarises this approach for polymer prints. For more detailed identifying parameters for polymer, metal, and other material prints, see the references tables in Chapter 7.

Step 1: Check the surface finish

For identifying processes used for polymer parts, the best starting point is the surface of the part. Run a fingernail across the part. If distinct lines can be felt on the surface, then it indicates the part was probably built using a filament extrusion process. If instead the surface has a slight grainy texture, but with little discernible "grain" direction on most surfaces, then it is likely to be a powder-based fusion technology.

In a dual-filament process, as shown in Figure 1.2, a soluble support material could be used as the second material and removed in a solution bath. This means there will not be any excess material left on the part where support material was broken off. In a single FFF process, there should be residue where the support structures were broken off, or evidence of finishing to clean them off the product. In addition, if the part is a functional part, rather than a prototype, then it is likely to have been made using an industry-standard dual FFF. It is harder to achieve a smoother and more consistent finish with a desktop single-filament printer than it is a dual FFF. Therefore, the higher the quality of the part, the more likely it is that it has been created using a dual FFF. If it has visible faults in the surface, it could be that the operator is not experienced, but it is far more likely to be a single FFF and that the problems are due to the settings and quality achievable on a low-cost printer.

1.2
Close-up detail of the layers on a dual-material 3D print produced on a desktop FFF machine using 0.2 mm layer thickness and a PLA material (grey) and TPU (white).

If the surface of the part is a mix of textures, for example, from flexible to rigid, then the part could be made by a material jetting process where there are a myriad of print heads (e.g., Stratasys J750) depositing and/or blending different materials much like a common inkjet colour printer. The part can be very detailed, but it tends to be made with materials that can deteriorate over time due to the photosensitive nature of the resin materials used, and it is unlikely to have good mechanical properties. These objects are generally used as communication tools, particularly for surgical planning. Adding texture as a 3D feature to an end-use part can provide additional functional advantages, such as added grip, or as a communication tool to help the user interact with the product. However, the addition of texture also masks the stair-stepping effect created by the successive layering processes used in additive manufacturing, which can be particularly visible on curved surfaces.

Step 2: Colour
Consistent with surface finish, colour can also be used to help determine the 3D print method. Colour 3D printing can be broken down into three main categories, supported by several technology categories:

1. Single colour – material extrusion, powder bed fusion, material jetting, vat photopolymerisation, binder jetting.
2. Multi-colour – material extrusion, powder bed fusion, material jetting.
3. Full colour – powder bed fusion, material jetting, binder jetting, sheet lamination.

Single-colour 3D printing is of course the most common and most straightforward process from a hardware perspective. If the visible layers described previously did not give it away, the use of a single colour other than white or grey for an entire part can be a sign that the part was most likely produced using FFF. Similarly, if the part has a grainy texture, if it is difficult to see layers, and if it is made up from a solid block of colour, it has most likely been produced using a powder bed fusion process (most commonly white for SLS, or grey for multi jet fusion, although they can be dyed).

If, however, the product has a very smooth surface finish and fine details yet consists of a single-coloured material, it will most likely have been made using a UV-cured resin process. This could be either material jetting or one of the vat photopolymerisation processes, both of which produce similar parts. Material jetting is expensive, relative to other processes, and is best used for extremely fine detail. It is accurate at a micron level, so it tends to be more useful for medical modelling than for products, and it typically prints using a secondary support material that can dissolve or melt away as part of post-processing. In contrast, vat photopolymerisation printers using stereolithography (SLA), as in the example in Figure 1.3, or digital light processing (DLP) methods, share similarity with FFF in that the supports are the same material as the part being printed. Translucency as well as solid colours are both possible with these resin processes, and the products can be suitable for end use applications, such as light shades and surgical guides (e.g., personalised drilling or cutting guides), although the process is most often used for visual models.

The plaster-based binder jetting produces solid white parts which feel rough and powdery. These parts will need to be sanded to create a smooth finish and remove loose powder. The material is plaster, and so the parts are suitable for visual models, more than anything else. The parts can be easily painted and varnished but cannot be dyed, as the material is not hygroscopic. Similarly, parts created using SLS are a single colour all the way through, usually white, because the powder is loaded from a hopper, and so parts cannot be multiple colours within a single print load. However, the material is hygroscopic, and so the parts absorb dye very well. This means the colour in SLS prints tends to be saturated.

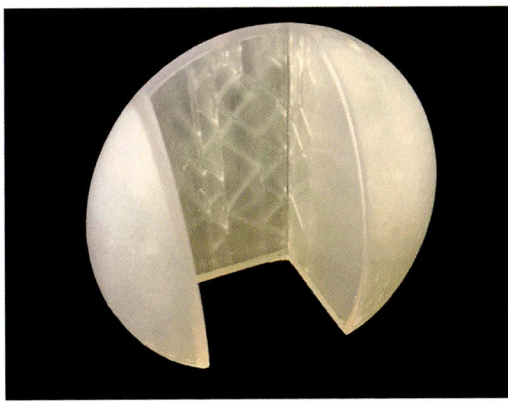

1.3
Stereolithography
(SLA) – laser-cured
resin with internal
resin support
structure (Loy).

White or strong single colours are therefore a characteristic of SLS prints. Black, however, may be indicative of HP Multi Jet Fusion technology (MJF), although it could also be SLS.

The first generation of HP MJF machines do not have colour capability but can be identified as different to SLS because of a colour characteristic caused by its printing process. Because the material is fused by a heating element rather than by a laser, the white powder has black dye added to it in droplets where the material needs to be fused. This allows the powder to absorb the heat, whilst around the edges of where fusion is specified, droplets of anti-fusion agent are added. This gives the parts crisply defined edges. It also means that the parts look dark grey in colour.

Multi-colour printing describes the ability to print with two or more different colours and/or materials simultaneously. This can be achieved with dual- or multiple-filament FFF machines, or add-on devices that help add this capability to a single-filament machine. Identifying these parts is similar to methods already described, except for the additional check for clearly defined blocks of colour, which may be up to four to five colours printed in the same part. These colours are not blended, so identifying boundaries between coloured areas, or slight bleeding as one filament was swapped to another through the extruder, can be clues.

Multi-colour printing is also possible with any of the full-colour print technologies. Full-colour printing, however, provides for the blending of different colours to create thousands of colour possibilities during the 3D print process. Much like an inkjet printer that has cyan, magenta, yellow, and black (CMYK) ink cartridges, similar cartridges can be used in some 3D print processes to add colour or blend between different materials. This can be as fine as a pixel, known as a "voxel" in 3D printing (volumetric pixel), with the print heads containing hundreds or even thousands of small nozzles that control the deposition and blending of appropriate materials or colours. Multi-material, full-coloured prints for visual models can be made this way using material jetting; but these prints tend to be used as communication tools, particularly for medical applications, as shown in Figure 1.4. Determining this print method from a part's appearance is relatively easy, as this is currently the only way to achieve full-colour multi-material parts that have a good surface finish, including translucency.

A second-generation HP MJF printer had complex colour mapping capabilities. These are discussed in detail later in the book, but overall, there are some indicators that let the product designer know a sample had been printed using this technology, as opposed to a material jetting process. First, the part will be a single material. This differentiates it from a material jetting technology where multiple materials can be combined into a single part alongside colour. It has a visibly grainy surface (when not post-processed with different surface finishing methods), like SLS parts. The colour definition is very good, similar to SLS when dyed. However, with MJF, complex images can be mapped onto the surface, with clear definition between the colours, as shown in Figure 1.5. This is not

1.4
Multiple material, full-colour material jetting technology used for medical models as communication and planning tools, such as on the Mimaki printer (a) and on a Connex (b).

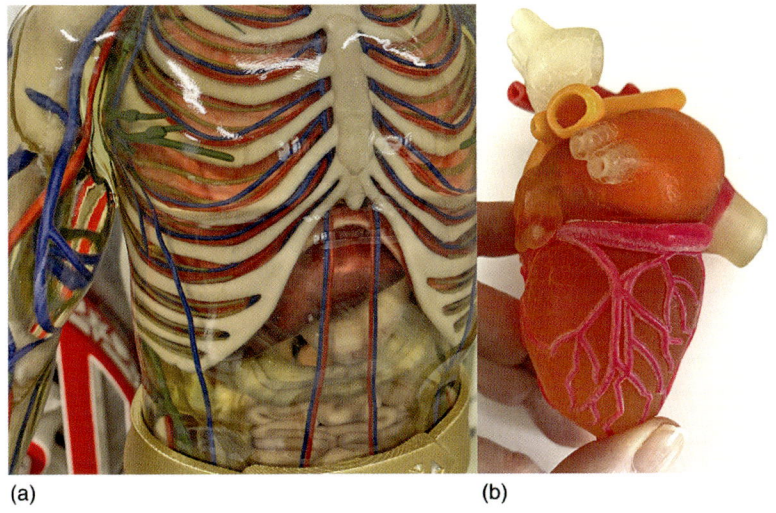

(a)   (b)

1.5
HP Multi Jet Fusion (MJF) colour sample that is cut open to show interior grey colour.

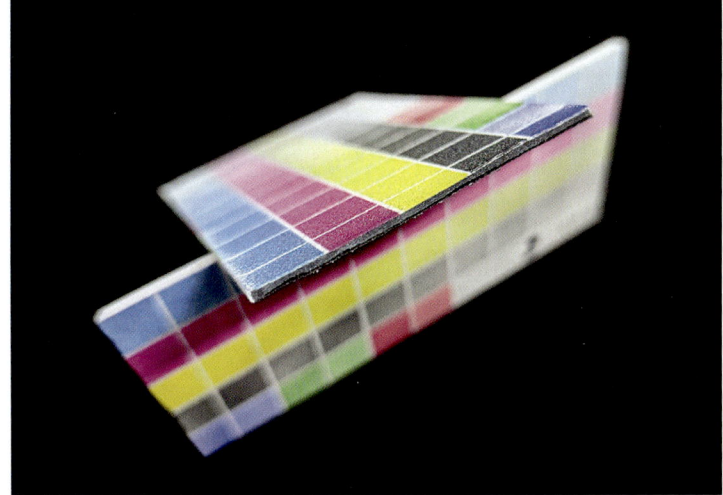

possible with SLS, where a whole part must be dyed. The colours can be very bright, or muted as specified, and are of high quality, as can be seen in the examples shown and discussed in the case studies in Chapter 6. If the part is cut open, it will be very clear that it is an HP MJF process. This is because the colour will be applied only to an external layer of white, while a black ink is used for the bulk of the part to aid with material fusion and results in a grey colour all the way through.

The points described here introduce an initial visual identification of processes that may be useful in evaluating printed parts or considering the use of different processes. However, just as with conventional manufacturing, there are strategies for evaluating whether a part is designed such that it can be manufactured using a process. A part also demonstrates designing for process and, beyond that, how to design to maximise that process. This information is provided as three levels in reference tables (good, better, best), with exemplars of best practice. These are created to help product designers build technical and design knowledge to support working with the industry engagement strategies, as outlined in Chapters 3, 4, and 5.

DESIGN RULES FOR SOFTWARE – WORKFLOW AND FILE PREPARATION

3D computer aided design (CAD) software has become one of the principal tools used by product designers over the last decade, with most graduates of university programs learning at least one CAD program during their studies in preparation for the workforce. Professional designers in industry may have completed online or face-to-face training to upskill as CAD has become more ubiquitous, and there are countless resources including books, video tutorials, Massive Open Online Courses (MOOC), and training offered by software companies. Assuming some expertise with CAD, this section will focus on some of the key considerations in going from CAD model to 3D printer, beginning with some of the basic requirements for those unfamiliar with 3D printing file formats, through to slicing files for 3D printing and more advanced strategies to improve and repair files (Figure 1.6).

1.6
The three main processes required to go from CAD to 3D print.

EXPORT

It may seem obvious, but additive manufacturing is a digital manufacturing process. Therefore, it requires a 3D digital file describing geometry to print it. Other manufacturing processes may not require a 3D file, instead being manufactured from traditional 2D drawings. For example, many sheet metal parts still require skilled metal workers to translate drawings into folded metal forms or may combine some digital processes like CNC cutting with more manual folding and welding processes. Designers must be familiar with the different file formats for 3D printing, including the advantages and disadvantages of each.

STL FILES

Almost every CAD program today facilitates the exporting of files suitable for 3D printing. The most common format, the Stereolithography (STL) file, is common in design studios, universities, and maker spaces, where it is used for prototyping. However, designers are less likely to be familiar with the more technical aspects of what a STL file is, and how the settings selected can dictate the success or failure of a 3D print before it even goes to the 3D printer. STL can also stand for Standard Tessellation Language, Standard Triangle Language, or Stereolithography Tessellation Language. A considerable amount of highly technical mathematical information can be found in journals for designers interested in this level of understanding [2]. However, in simple terms, a STL file is a list of triangular faces that, when visualised, forms what designers would know as a *mesh* file, particularly familiar to those with experience with animation and sculpting software. As a mesh, the STL file only describes the surface geometry of a part; it is not solid, contains no dimensional or colour information, and, as a result, can be very difficult to manipulate in any meaningful, accurate way. It was developed in the late 1980s, alongside the earliest stereolithography 3D print technology, and is in many ways an outdated format, as might be expected. Yet, it remains the most common file format for 3D printing despite new formats emerging, as will be discussed following this section.

The basic process to create a STL file is straightforward for CAD users. In almost all software, a designer simply needs to use the *Save As* or *Export* functions and select STL as the file format. Depending on the CAD software, the user may be presented with several settings to modify, the most common being the option to choose either *binary* or *ASCII* as the output method. Without getting into technical details which do not concern most product designers, these are simply two different methods for describing each triangular face that makes up a mesh and storing these as a list of instructions. The most popular choice is to use the binary method, as it results in smaller file sizes, and this is perhaps the only thing most designers will need to know.

Other settings important for designers to consider will be in the options associated with how accurately the triangular faces represent the original CAD geometry. Because most product designers will be using CAD software that precisely represents surface geometry, converting to STL format will alter the geometry as a file goes from solid/surface geometry to triangulated mesh. For those familiar with 2D graphics, it is like going from vector file (e.g., Adobe Illustrator) to raster/pixel file (e.g., Adobe Photoshop). As a result, STL files are described as a *lossy* file format, with information lost during the conversion process. Exporting to STL format becomes the final step in the CAD process, an output that is separate from the critical CAD file.

Some CAD software may provide fixed options for the quality of the STL file, while others will allow greater customisation of this information. Figure 1.7 illustrates the conversion of a CAD model of a tube into both low-resolution (low number of triangles) and high-resolution (high number of triangles) STL files. The tube is no longer made up of perfectly smooth surfaces in either type of STL files. However, the high-resolution file clearly provides more detail, with finer planar triangles that more closely represent the original CAD model. The low-resolution file becomes more faceted, with larger flat triangles used to represent curved surfaces. In an extreme case, the circular CAD model would become more octagonal as the curved surface broke down into larger and larger

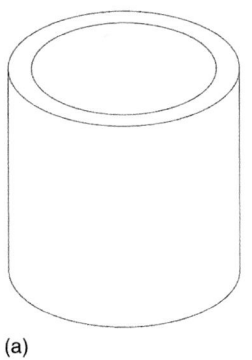 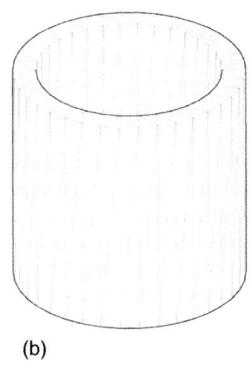 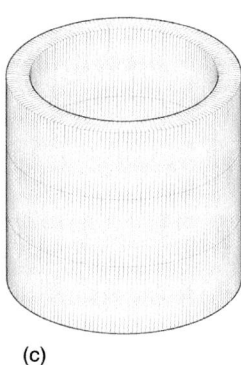

(a)   (b)   (c)

1.7
A solid model of a tube in CAD software (a) compared to a low-resolution STL (b) and high-resolution STL (c).

Table 1.1 Data for tube STL files

Tube STL file	File size	No. triangles
Low Resolution	15 KB	288
High Resolution	126 KB	2560

triangles. This is rarely desirable where a product has been carefully designed in CAD and requires high accuracy, although low-resolution settings can be used for more artistic effects where the triangulated surfaces become a feature.

Other than surface quality, there are other reasons these settings matter for a designer. As an example, key pieces of data about the tube STL files are shown in Table 1.1. In this case, the low-resolution tube is made up of 288 triangles, resulting in a file size of 15 KB. In comparison, the high-resolution file has almost ten times as many triangles at 2560, resulting in a larger file size of 126 KB. There is a direct relationship between the number of triangles and file size, which is not important for a small tube but becomes critical for larger, more complex forms which may be made up of millions of triangles. Small changes in STL export settings not only significantly vary the accuracy of the geometry, but they also affect file size, which can make any future processing of the STL file for 3D printing difficult, or even prohibit 3D printing due to limitations on file size for some software/3D printers. Therefore, a balance must be found based on the accuracy required, as well as the file size for a specific part, and it is recommended that designers experiment with settings as part of their additive manufacturing journey.

Colour files

As STL files do not provide colour information, there are additional file formats for complex colour models suitable for 3D printing. There are two popular formats for 3D printing colour that use jetting processes, such as material jetting and binder jetting, or multi jet fusion. The first is the Wavefront Object (OBJ) file, which can be thought of as similar to a STL file; it is a mesh made up of

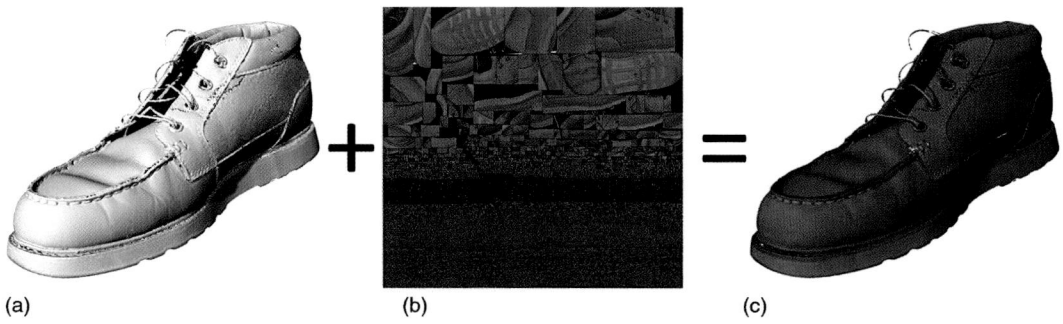

(a)   (b)   (c)

1.8
The two core components of an OBJ file are the 3D mesh file (a) and PNG image file (b) that combine into a full-colour model (c).

triangular faces with separate colour or texture information due to its foundation as an animation file standard. While less common than STL files, this format is available as an export setting in almost all CAD software and is particularly common for files that have come from a 3D scanner, translating colour information in a universal file format. As shown in Figure 1.8, an OBJ file is actually a set of three files: one describing the external mesh geometry of an object, another an image file (PNG) containing colour and transparency information to be stitched onto the geometry, and the third being a MTL file that describes how the image is wrapped over the mesh.

The second popular option is to use a Virtual Reality Modelling Language (VRML) file, or the more updated X3D file. While technically different to the OBJ format, the principle is the same in that there are separate 3D geometry and colour files, and both are required for colour 3D printing. VRML files are popular for web applications, as they also contain rendering information, such as reflections, as well as animation and interactive information. The OBJ file format has been the most popular for colour files and is supported across a broader range of software and 3D printing platforms than VRML or X3D. Both file formats can be 3D printed even if there is no access to a full-colour 3D printer, for example, on a single-nozzle FFF machine.

New standard file formats

It is interesting to consider why conventional 3D printing file formats can appear so outdated. It seems unreasonable to convert a perfect CAD model into mesh files and start to lose quality in the geometry. This is not required for most other manufacturing processes, especially digital CNC processes. Another question may be that if the designer is more advanced in the use of 3D printers, how can different materials be specified throughout the part, rather than purely as external colours? This is where new file formats have been developed and are increasingly being adopted by CAD software companies, slicing software companies, and 3D printer manufacturers. It is a rapidly developing area and will be particularly relevant for product designers in the future.

The 3MF Consortium has led the development of a new open-source file format known as 3MF, capable of containing all the necessary information about a part for 3D printing within a single file, including geometry, colour, materials, and other properties. This file type has been led by industry. Another file format called the Additive Manufacturing File Format (AMF), providing a similar advanced file capable of containing all necessary information about a part for additive manufacturing, has also been defined within ISO/ASTM 52915. One of the most important aspects of these file formats for product designers is that it allows them to control exactly how a part will be produced, rather than simply exporting "dumb" geometry, and having to separately slice it for 3D printing on a specific machine or allow a manufacturer to select the process parameters. All these settings can be contained with a 3MF or AMF file, ensuring that the designer is in control of the production process no matter where the part is printed, or even on what machine it is printed. While settings can be changed by anyone with the file – for example, a manufacturer who may have greater knowledge of the settings optimised for their specific machine – these new formats allow the designer to better communicate intent, as they have done with other manufacturing technologies.

The new file formats also allow for better communication of complex structures like lattices without resulting in large files, which in STL formats add significant file size due to the number of triangles required to define the topology of the lattice. This means that more complex files can be created and readily 3D printed or modified, removing some of the limitations that have constrained designers over the last few decades. While not as universal as STL or OBJ files, most professional CAD software will accommodate for these file formats in recent versions, as will many slicing programs and dedicated 3D printing software. It is likely that one of the new file formats will become the new universal standard, replacing the need for any other file format except for older machines running older slicing protocols.

Check/repair

Professional product designers usually use primary CAD tools that are solid modelling programs, such as Solidworks or Fusion 360. It is very unlikely that errors will be encountered when exporting 3D printing files from such programs due to the way geometry is built, checked, and constrained in such software before even being exported. Therefore, checking and repairing of files may be unnecessary, though it is a good habit for the designer to practice, especially before sending a file to a bureau or manufacturer. Errors in geometry are far more likely in more flexible environments, such as those in Rhinoceros or SketchUp, where intersecting forms, and different surface and mesh modelling methods, are combined. Errors are even more likely if the STL or OBJ file has been imported from an online source, or there have been any direct modification to such a file, rather than to the original CAD data. This includes direct mesh modelling in software such as Blender or ZBrush, or the use of 3D scan files.

The golden rule for 3D printing files is they must be *manifold*, also known as *watertight*. This means that the mesh must have no gaps, cracks, or missing

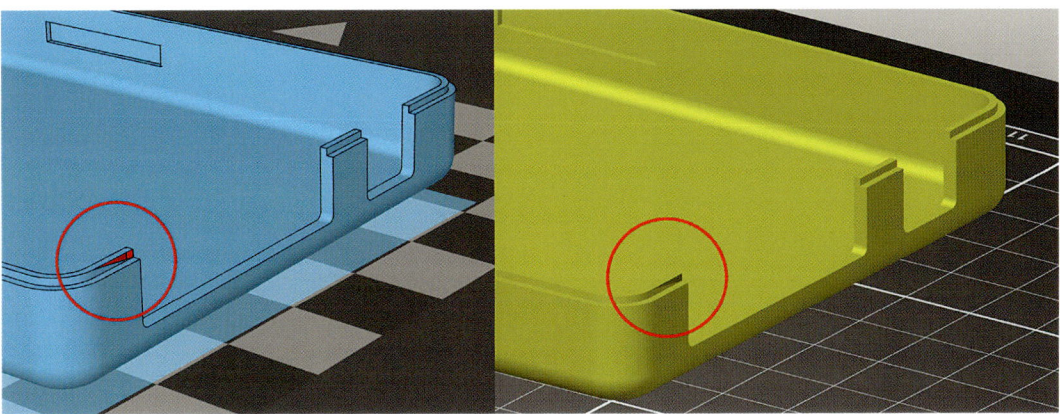

1.9
The same STL file with error in two different slicing programs, with only one of them clearly indicating a problem through red colouring.

triangles that could hypothetically "leak" water if they were filled up like containers. If anything has gone wrong with the conversion of solid model to mesh, or the file was not properly modelled to begin with, errors are likely [3]. Just one error can be enough to cause a file to fail. While sometimes an error will be automatically picked up when the part is "sliced" (see Chapter 7 for more details) – that is, when the part is split into layers ready for printing – many slicers will attempt to correct or ignore the problem, which is an issue for the designer, as this will give the impression that the part is fine, until it starts printing and there are problems. For example, Figure 1.9 shows the same STL file in two different slicing programs. In one of them, the error is clearly shown by differentiated colour (a single missing triangle), while the other is very difficult to notice. In both cases, the error is not significant enough to prevent 3D printing; however, the area with the problem may not print as desired and must be corrected. If this were a larger missing triangle, or multiple triangles, the chances of failed printing would increase.

Option 1: Fix the source file
Whenever possible, it is best to correct the original CAD data, or modify the export settings, rather than try to repair a damaged 3D print file. This ensures the 3D print file is the most accurate representation of the design intent, as repair methods for 3D print files can change the geometry of a file, especially automated methods that a manufacturer or print bureau might employ where they do not know what the part is for and may not even look at the file.

If an error has been picked up during slicing or using an additional program that checks 3D print files (e.g., the free Autodesk Meshmixer program), it should be returned into the CAD program. The geometry that is affected can then be modified on the original model. This may be as simple as increasing a fillet radius, or a more complex process of re-ordering the way a piece of geometry was constructed in the affected zone. As an early preventative measure, many CAD programs will include built-in tools to check models before even exporting to a 3D print file.

> **DESIGNER'S COMMENT**
>
> Always fillet (round) internal corners when working with additive manu-facturing. Sharp internal corners can become a source of stress concen-tration and weakness on your part. It never hurts to fillet external corners too. It reduces the amount of material used (and therefore the cost) and makes the product more comfortable to hold.

Option 2: Export with different settings

If it is not obvious where the error is occurring in the geometry, another option is to try exporting the file again using slightly different settings. As discussed earlier, converting from perfect solid geometry to a triangulated mesh involves losing information, and in this process, it is possible that some geometry con-verts incorrectly. File conversion is a bit of a "black box" technology, and some-times the adjustment of parameters can change the result for a positive outcome.

Option 3: Repair software

An essential piece of software for product designers involved in additive manu-facturing is one that provides some form of mesh repair. Repair tools are availa-ble in some CAD software used by designers like Rhinoceros; however, dedicated programs for 3D printing are also available to streamline the process and provide dedicated tools to prepare files. Numerous programs exist, from free (e.g., Autodesk Meshmixer) to paid (e.g., Materialise Magics). They will typically include automatic and manual methods to repair files. For example, Figure 1.10 shows part of the automatic process in Materialise 3-Matic software

1.10
Mesh repair software can be used to highlight and correct problems in 3D print files, in this case overlapping triangles.

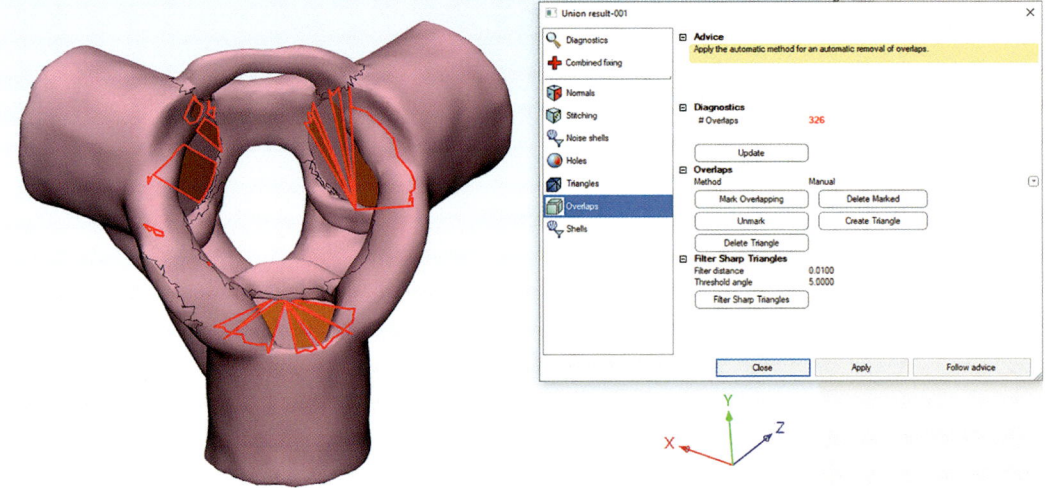

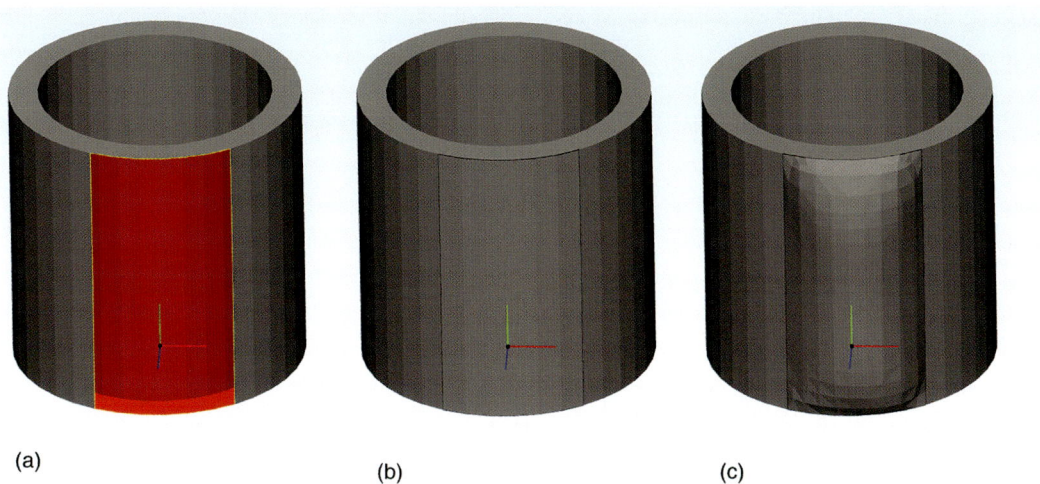

(a)   (b)   (c)

1.11
A mesh hole (a) which has been repaired to follow surrounding geometry (b) compared with a more organic addition (c) that does not suit the designers' intent.

to fix problems, which in this case lists 326 overlapping triangles. For such a feature, deleting unnecessary triangles that are causing problems can result in a better file for 3D printing. Other software tools are available to address: crack repair (where the edges of triangles are not perfectly aligned); inverted normal (where the face of a mesh triangle is flipped the wrong way, with the outside surface facing to the inside); hole repair (filling in missing triangles); and the removal of small meshes (aka noise shells) that may be hiding within the main mesh or somewhere around it.

In most cases, automatic tools will quickly and accurately repair problems. However, in some cases the result may not be desirable. For example, Figure 1.11 shows a mesh hole that has been repaired using two different methods, resulting in very different geometries – manual and automatic. Automatic methods may sometimes result in an undesirable result, and a designer must manually repair features. Depending on the repair software, and the complexity of the problem, this may be easy or may require a careful process of rebuilding a section of a model triangle-by-triangle. Therefore, modifying the original CAD file is preferable to repairing a 3D print file, which is why designers must be responsible for the repair of files.

Slicing

The fundamental characteristic of additive manufacturing is that parts are manufactured layer-by-layer. Therefore, 3D geometry must be converted into individual layers for the machine to produce. This is known as *slicing*. Slicing software is used to set up the process parameters to produce the part, or multiple parts, and includes vital information specific to the 3D printer (e.g., build volume, temperature, and material) as well as information that has some flexibility and can directly affect the quality of the outcome (e.g., print speed, layer thickness, and infill structures) [4].

1.12
Typical file
conversion stages
from CAD to 3D
printer.

New file formats like 3MF and AMF can contain this information within them already, whereas STL and OBJ files must be sliced separately, with settings applicable only to a single type of 3D printer. This process turns a 3D model into a series of machine instructions known as G-code, which is used by the 3D printer to control the production of parts step-by-step. As shown in Figure 1.12, this means that the initial CAD file has typically undergone several conversions by the time it is 3D printed, especially when using the common STL/OBJ formats. It is inevitable that problems and inaccuracies occur.

Many designers who have been using desktop 3D printers will have a good understanding of the links between what was designed in CAD, the settings chosen during conversion to mesh, slicing process parameters, and the resulting print. A perfect STL file will still fail to print successfully if the wrong settings are selected during slicing, perhaps failing to use support material in the right locations, printing too fast, or not printing at the right temperature, causing materials to clog or not fuse correctly. Similarly, a part designed specifically for one production process, like SLS, with fine lattice details may not print well using FFF. The processes are very different, and the same results cannot be expected just because a part is designed for 3D printing. Designers need to be specific, and it is important to state whether the model has been designed for SLS, as opposed to MJF or another completely different 3D print technology, rather than being a general design. It makes a difference, and it will elevate the work as well as product design discipline in design for additive manufacturing.

Depending on the type of 3D print process, as well as part complexity, slicing can be a quick process requiring a couple of minutes to orient a part and check settings, or it may take a significant amount of time, especially within a manufacturing context with many parts printed as a single batch. As shown in Figure 1.13, the process begins with the imported 3D print model. Information related to the production of the part, including layer thickness, print temperature, print speed, support material, and many other factors, are modified. The model is then sliced into layers using these settings. Figure 1.13 shows a sliced part for FFF production, and the detailed image on the right shows the resulting pathways the nozzle will travel to produce some of the layers. In this case, the red colour is used to represent external walls, green is used for internal solid fill, and the grey layers are showing which layers occur below the current layer being visualised.

As a designer advances to some of the more complex polymer processes that provide multi-material printing, more time is required to specify material

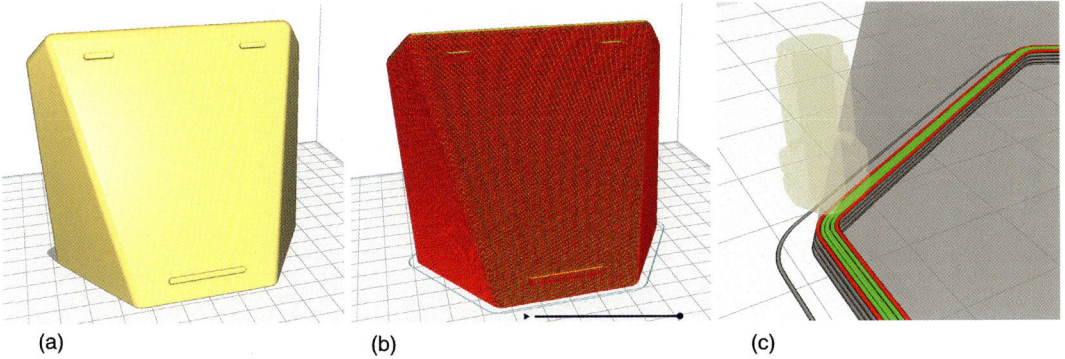

(a)   (b)   (c)

1.13
Imported STL (a), sliced model (b), and detail of some of the slices showing the pathway of the print nozzle (c).

properties and set up colour information. Metal processes also require significant setup time, with requirements for supports to distribute heat and anchor parts to the build plate. Some automatic tools are available; however, to maximise the capabilities of such printers, manual setup remains essential to metal 3D printing today. As shown in Figure 1.14, this includes creating angled support structures (e.g., yellow section) that will allow for easier post-processing compared to supports that intersect with the part being printed that may have been added by the software automatically.

Designing for additive manufacturing processes requires a designer to understand what will be required during production, such as support structures in SLM, and attempt to design the part to require a minimal amount of supports or accommodate areas to be supported without the support structure intersecting the part. As well as the potential to improve surface quality, for example, the time taken to set up a part for printing is a cost that manufacturers will factor into the price per part for production, as is the extra cost of material used for supports and later removal. By understanding the slicing process, and the

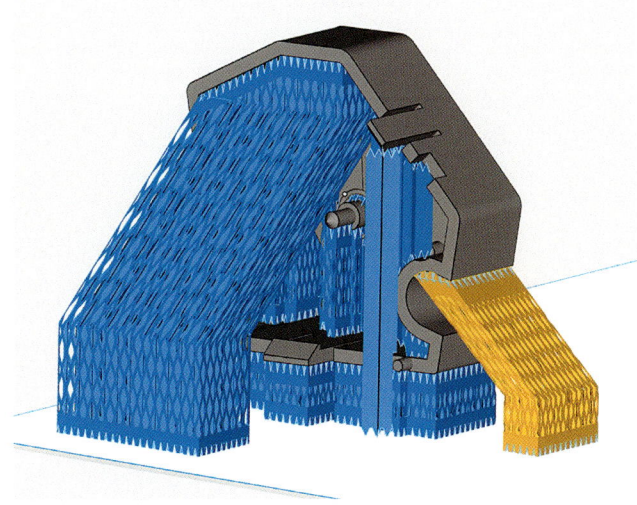

1.14
Setting up different supports for a Selective Laser Melting (SLM) process.

different requirements of the varied additive processes, a designer can have a significant impact upon the quality, cost, production speed, and post-processing requirements.

GETTING HANDS-ON WITH HIGH-END INDUSTRIAL POLYMER PRINTING: KEY STAGES OF THE WORKFLOW DESIGNERS NEED TO KNOW

Product designers are engaged to create new outcomes. One of the reasons this can be required is when a company invests in a new technology. Faced with an industry machine for the first time, whatever the technology, two essentials will need to be researched and understood prior to any design development work. These are:

1. The basic constraints of the technology that must be adhered to for a product to be able to be manufactured.
2. The opportunities provided by the technology to exploit its characteristics and justify the investment in the technology, and the operation of the technology.

For the first requirement, the designer must research the technical and sales information provided on the technology to identify what is pertinent and reliable as the basis for design decisions. To understand the second of these requirements, the designer can model and print test pieces that illustrate what can be done, what works well, and what fails. For product designers new to working with 3D printing technology for functional products in an industry or design consultancy setting, these approaches apply as much for these technologies as for traditional manufacturing ones. The following sections provide an insight into the realities of working with selective laser sintering technology. Some of this information is discussed in detail in the engineering literature; however, some of it is poorly documented and relates to workflows and experiences that are learned over time. This is often not captured in glossy sales brochures.

Introduction to a selective laser sintering 3D printer

As outlined earlier in the chapter, there are a myriad of 3D printing technologies, and for each one a designer will need to sort out the key points that will be needed to work effectively with each one. This section provides an insight into the practicalities of hands-on 3D printing with one of the most common types of polymer 3D printing technology, selective laser sintering, despite the process being applicable to any new 3D print technology encountered.

The EOS Formiga P 110 SLS printer was chosen for this example because it has a proven track record in industry, and because for the product designer, it is typical of the type encountered in industry at this time. Numerous companies are producing selective laser sintering machines, so there will be differences in the build envelope, and in the practicalities of the machine. For example, the feed mechanism may be different. One of the reasons for choosing this machine is it can be operated by a single user if required. It has a relatively small footprint, and therefore, it can be installed in a relatively small area. Additionally, it has a proven track record. However, this is not a recommendation of this machine over others that are similar because the most suitable machine for a company will depend on the specific needs of that company as they relate to details such as build envelope, laser strength and quantity, build space and heating element configuration, feed delivery system, and so on. Nevertheless, this example provides a good starting point for understanding working with the technology as a designer.

Another reason this is a good starting technology is that no vacuum is required in the build space, meaning it is accessible to use. No argon gas is required either. Argon gas is dangerous if inhaled and therefore needs to be used in regulated conditions. Nitrogen is used in the build space of the EOS Formiga P 110 SLS, to help damp down the powder, but this is not a hazardous gas. These types of printers would be good introductory machines for design studios and universities. Yet, even though they are increasingly prevalent in engineering environments, they are still not common in product design facilities and design studios. Arguably, this is because the technology came out of engineering, and therefore, disciplinary leaders in product design can lack the experience or confidence to bring the technology into the department, particularly because of the financial investment required. The step-by-step description of hands-on working with an SLS printer is intended to demystify the process and provide insight and confidence for product designers and academics looking to work with the technology.

Starting point: The build envelope

As an entry-level selective laser sintering machine (this type and size of SLS generally cost less than US$140,000/120,000 pounds sterling in 2020, depending on capabilities), the build space, or build envelope, is the key point that a designer needs to know before starting. The build space is the usable space within the block of powder that can be filled with parts. For example, the build envelope of the EOS P 110 is 200 × 250 × 330 mm. However, because the plastic shrinks, the prints must be scaled up automatically in the software to compensate, and therefore, the actual build size of the box is slightly smaller at 193 × 242 × 321 mm.

In addition, the base of the build space and the thermometer in the build space (indicated by a red box in the accompanying slicing software) both typically have around 50 layers of powder around them (approx. 5 mm). The base layer acts as an insulator to prevent the model from being distorted by a temperature fluctuation from the base plate and placing parts where the temperature of the build chamber is measured, as this can affect the accuracy of the readings and, therefore, the overall temperature of the build. As a result, when a designer first designs for a machine that technically has a build space of 200 × 250 × 330 mm, when starting, it is recommended that parts are kept to under 180 × 230 × 300 mm to be safe. When the machine is new, the user prints calibration pieces to check the shrinkage, but once the machine is set, that is usually fixed. In addition, as per the description of file preparation, gaps will be needed around the parts to prevent them from adhering to each other during the print. Fusion is achieved by laser sintering, and material around the part can be affected by the overrun of the laser. It is not a precise action, and therefore, each part should have an insulating layer between it and the surrounding parts.

The EOS Formiga P 110 has a 100/150-watt laser. The strength of the lasers in a machine, and the number of them, will affect the print time. The stronger the laser, the faster it can heat and fuse the powder. Additionally, the more lasers there are, the faster each layer of the print can be completed, as each laser works to complete a portion of the scanning instead of a single laser which must fully scan each layer by itself. After each layer, the container drops by 0.1 mm, and a new layer of powder is spread across from the left or right (alternating). Printing a solid block of the size of the build space (which would make no sense to print) would take an estimated two to three days, given the speed of the laser and the layer thickness in this instance. In general, for full-size, selective laser sintering machines, a full height print of approximately 300 mm would take over 20 hours to print. An average print of parts in the recommended 8 percent fill in a basin will take between 6 and 8 hours. However, it will then take around the same time for the machine to cool down before the parts can be removed. If this is not done, then the parts could warp. It is possible to remove parts during the cooling process for advanced design applications to exploit the tension created in the part by doing this. However, as a starting point, this is not recommended.

Preheating

To start the build, the chamber must be preheated to close to the temperature of the laser. For example, where the laser melting temperature is 200°C, the build chamber temperature is set at 165.5°C. This is not fixed, and for advanced applications this could be altered; but in the majority of uses alteration is not required, and the temperature remains set. The printing element providing heat in an EOS Formiga P 110 is straightforward. It must be regularly checked, as any problems with it can cause uneven heating and, therefore, an uneven build. In addition to the build chamber being heated, the removal chamber (in this case, below the process chamber), where the hopper is, also needs to be heated. By heating this chamber to around 150°C, the prints are less likely to distort as they are lowered with each new layer; otherwise, they would be in powder that was hot at the top and cool as they were lowered towards the base. Preheating takes around 1.5 hours for this machine. In addition, time must be allowed to unload the hopper, remove the prints from the powder, and clean the individual pieces, which takes some time. This means in total a full build will take around 20 hours, even without file preparation and setting up the machine with the parts.

Realistic planning must take this into account, and a single build needs to be maximised. Partial builds are possible where the powder is layered over the part only as deeply as required to ensure a consistent build, which will reduce the actual build time but have little effect on pre-heating and cool-down times. Therefore, maximising the quantity of parts that are printed and filling the build envelope each time is ideal to improve efficiency.

KEY POINTS

- Build envelope is the chamber specified in the technical information provided by the company, less an allowance for the automatic scaling up of parts to compensate for shrinkage and to insulate the base, sides, and heating element in the chamber, as well as each part, from surrounding parts. Reducing the available build space for the design of parts by approximately 5 percent would be a good starting point for a designer working with the technology for the first time.
- A full build of mixed parts will take around 1.5 hours to preheat and could take up to 20 hours to build, but this will depend greatly on the intensity of the build design. For example, a highly detailed 3D fabric design will take the maximum, whereas one that has a high material volume but involves less sintering could be much quicker to print.
- The cool-down time is recommended as approximately half the build time, but realistically for a 6- to 8-hour build, then allow 6–8 hours for cool down, plus the time to extract and clean the parts. Overall, allowing for a minimum of 20 hours for a build (and as much as 30 hours for a more detailed, full build) gives the designer a realistic expectation of when a part could be ready after it has been loaded into the printer software.

Powder

The unused portion of powder that encased and supported the parts is collected to be used again. However, it is important to note that the material has been heated, both by the chamber heater and from over-run off the laser which ages the powder, and this affects its performance characteristics. If the same powder was re-used on its own, the prints would not have the same integrity. This will result in problems of delamination between the layers, where the sintered material fails to adhere properly to the surrounding layers, and mottling, which leads to a surface much like that of an orange peel. This powder could potentially still be used for test pieces, as it does not damage the machine, but the parts will not be as strong. Based on detailed mechanical testing, the machine and material suppliers provide recommendations on how much of the powder can be reused if upgraded with fresh virgin powder. For example, with an HP MJF printer, the recommendation is approximately 80 percent used powder and 20 percent new powder.

3D printing workflow

Once the files have been imported into the machine's software, there are parameters that can be tailored for the parts being printed, or material being used. For example, the strength of the laser can be tuned, or a simulation of the tool path can be run to check for any conflicts and identifying where parts may overlap, processes which can sometimes be difficult for the operator to identify amongst a large collection of parts "floating" in a 3D build envelope. The simulation runs line by line, and a clash will be highlighted by the software. It is possible to adjust the positioning of the parts to correct this type of issue. Small pieces can also be placed inside a build cage, so they are not lost during the removal and cleaning processes. Once the file is correct, it can be saved again in the software and resubmitted to the machine for 3D printing.

The first 50 layers build an insulation layer and are therefore blank. Then the print layers start to appear. The printer can start printing whilst the files are still loading because of this insulating gap. At this point there is no actual printing – just the wiper blade moving across the build plate, spreading a layer of powder on each pass. The wipers spread powder across from right to left, and then from left to right.

In the EOS Formiga P 110, the laser traces the outside edges of each layer first, followed by the inside detail. This is not the case for all printers. In advanced applications, the order of the tool path can have an impact on the accuracy of a very detailed part, and the designer will need to check the output through test prints and liaison with the technical officers and, if necessary, the machine suppliers. The level of control over printing order may be limited by the software available for a printer. For most production parts, though, this is not a major concern. The printing stops when the layers with prints in it are completed (i.e., it does not need to print the entire 330 mm build height).

Post-printing

Once the machine has cooled, the print chamber is lowered and excess material vacuumed out of the build space. The cake of powder in the square hopper shown in Figure 1.15 is taken to a de-powdering station. The tray can be pushed up to release the block of powder. The parts are manually removed from the powder, and then they are usually blasted with old powder to remove excess powder. This is relatively safe, even for delicate parts. Unfortunately, the powder collected from this process is not suitable for re-use because the powder that has adhered to the part tends to have been affected by the laser, causing it to stick. If the powder remains too difficult to remove (as illustrated in the example in Figure 1.16), glass bead blasting can be used. This is more damaging to the surface of the parts, so it should be used with caution, but it can be necessary where powder cannot be removed with polymer powder blasting.

1.15
Process of removing the "cake" from the SLS printer and then removing parts by hand on a table with a built-in vacuum to collect powder.

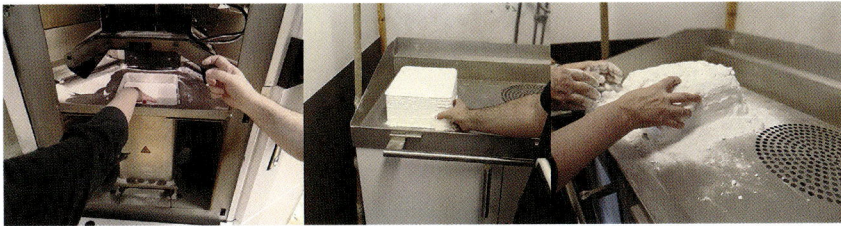

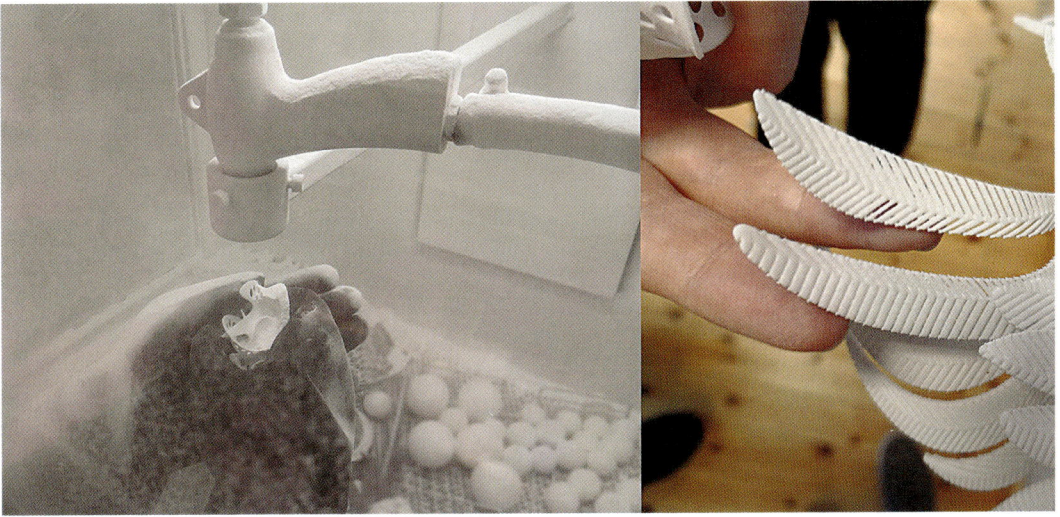

1.16
Blast-cleaning the parts and close-up of a part with fine details where blasting does not completely remove excess powder.

Removal of the powder provides designers with one of the constraints of working with this technology. In an assembly, small parts can have a tolerance of 0.3 mm. However, larger parts absorb more heat and, therefore, need larger gaps; generally, 0.4–0.5 mm is recommended. As a starting point, leaving 0.5 mm will be a safe tolerance, and it will also meet the design requirements for online service providers. As the designer works more frequently with the technology, tolerances can be modified within a model for specific design requirements, based on experience and testing. However, the automatic evaluation software of online service providers may reject smaller tolerances. If this is the case, then contacting the service provider to discuss the reason for a smaller tolerance and providing data on test pieces will help. In general, online service bureaus are open to discussion on changes for a customer's specific needs, particularly where the designer can demonstrate the specifications based on testing.

The standard layer thickness is 0.1 mm. This cannot be changed in the standard machine, as it is linked to the size of powder particles and to the ability to spread an even layer of these. It is possible to buy an add-on that allows for thinner layers, but there is little added benefit, as it makes the print time slower, with little additional quality. However, it is possible that in gently curving, long, flat, or reasonably flat shapes, lines can be seen, but for any curve over 10 degrees, no lines are visible without a microscope. This gives selective laser sintered prints their characteristic lack of grain direction. The other problem with thick, flat parts is that they tend to distort and so should be avoided. Generally, thick, flat parts can be made using other technologies, so they are of arguably poor design for selective laser sintering.

Whilst thick, flat objects are a problem, so too are very fine, tall prints. This is because of the mechanical action of the wiper blade. As the blade passes across a very fine vertical feature, it will tend to push the feature sideways, and fine features will not have much strength to resist the push of the blade. This can often be the cause of broken or failed prints. Blades can be made of different materials in different machines, so some wiper blades are more damaging than others. However, even wipers made from flexible blades will damage very delicate vertical parts, and so these design features should also be avoided or oriented differently where possible. It is recommended that vertical features be a minimum of 1 mm in thickness. If features do not print, there will be a problem not only at that point but also in the next layers. The printer will often continue despite a small broken detail, although the part quality will be compromised, or else the machine will crash, voiding the entire build from that height upwards. Avoiding this is critical for the designer, as not only will their parts be affected, but it could also affect the entire build. When starting to work with the technology, err on the side of caution and avoid chunky, flat pieces and very thin vertical stacks. If visible lines are seen on the surface of test prints, increase the angle of the parts or change the build orientation.

KEY POINTS

- Selective laser sintering with polyamide, supported by the surrounding powder, allows for organic shapes to be created with complex, compound curves.
- Because the powder acts as a support material, complex, multi-layered lattice structures can be created without any concern over removing supports.
- Access for the removal of powder is a constraint, and if the powder needs to be removed using blasting, then that powder will not be able to be reclaimed.
- Selective laser sintering with polyamide allows assemblies to be printed from a single STL file. Tolerances are a constraint, as powder can become trapped and difficult to remove, which will prevent the parts from moving.
- The new and old powders must be thoroughly mixed prior to use, which can be achieved by mounting the hopper onto a roller to tumble the contents. A feeder fitting can then be added, replacing the cap, and the hopper fitted back, neck down, into the top of the machine. This is not difficult for one person to do.

Identifying problems

If there are problems with the test prints, check for one of the following:

- *Overheating*: Overheating the powder (e.g., if the heating element remains on too long or is too hot), results in the powder fusing into a more solid block than it should be. This is not the fault of the design; rather, it is a technical issue that needs to be corrected.
- *Burning from the cleaning process*: In this case the object was held too close to the glass bead blaster nozzle for too long, resulting in the blackened surface on the object.
- *For assemblies that are solid*: For example, where a joint will not flex, it is likely to be a tolerance problem. Ensure that a minimum of 0.5 mm tolerance has been specified for initial testing, and then reduce the tolerance from there down towards 0.3 mm if a tight fit is needed to find the optimal tolerance. The size of parts will affect the tolerance as larger parts absorb more heat and will require larger tolerances.
- *If the parts have a mottled surface*: Or, if the layers are not properly adhered, this could be due to the mix of old and new powder. This is not a design issue.
- *If the build crashes*: This could be due to part files overlapping, which is a file preparation issue; or it could be due to very fine parts that have not printed properly because they are too delicate for the wiper blade; or it could be due to a solid part distorting during the build due to the amount of heat it absorbs. In either case it is a design and/or preparation issue. If testing the limits of printing for a design, ensure that parts that could crash are placed at the top of a build envelope, so that if there is a problem, it will not affect the build below, and other parts can still be salvaged.
- *Solid parts*: These can be an indication that insufficient holes were designed into the part to allow the powder to be drained out of hollow areas. These need to be designed into the part. If, however, the design requires loose powder to remain encased in the part, it is advisable to notify those directly involved with the printer, whether in house, or in a service bureau, as they will frequently add holes to release the powder automatically. This is because the cost is determined by the powder used.

This description of hands-on working with an industry-level, polymer powder-based selective laser sintering machine aims to demystify the technology for the product designer, product design student, and design academic. This includes designers who may have utilised SLS through a service bureau or supplier but not actually set up or post-processed a print themselves. The next section aims to do the same for metal printing. It introduces a practical, hands-on approach to working with metal printing, explaining the process step-by-step and pointing out the realities of working with the technology, through a designer-maker example of bespoke product design.

AN INTRODUCTION TO METAL 3D PRINTING FOR PRODUCT DESIGNERS

The current emphasis in metal printing is on replicating, as far as possible, the performance characteristics of existing parts created using conventional technology. Since 2012, with the 3D printed LEAP fuel nozzle by GE Additive, there has been growing interest in engineering in product redesign. This work has focused on three main areas: topology optimisation, generative design, and part consolidation. For product designers, the replication of existing parts is of little interest as they work on design development and innovation. The replication of parts already made for mass production using additive manufacturing would be of use only where a part could not be manufactured conventionally but had to stay the same geometrically, and in terms of performance. This would rarely be the case. Exceptions could be for spare parts, or for a part that is required as a one-off at a different scale to the mass-produced item.

For the product designer, new parts not previously possible with conventional manufacturing are of greater interest. This could still involve utilising engineering techniques, such as topology optimisation and part consolidation. However, product design, as a discipline, is where innovations in application design are most likely to occur. For this reason, the product designer needs to know the realities of working with the different technologies more than for most disciplines as they are most likely to stray from the conventions and guidelines established by the engineers who built the machines.

Use of service bureaus

For a product designer, starting to work with metal additive manufacturing can be particularly challenging. This is because, unlike working with polymer processes, it is difficult to work up through similar processes due to cost limitations of both machine and material. As a result, one of the approaches a product designer may use is to work with an online service provider for their initial metal prints. The advantages of this are:

- *Pay only for what is printed*: There is no upfront cost of purchasing a machine, and no steep learning curve to simply operate it.
- *Access the latest technology*: Most 3D printing bureaus amortise the cost of their machine quite quickly, and therefore can afford to remain updated with the latest machines.
- *Gain feedback*: Many online service bureaus incorporate automated tools to check files for their ability to be 3D printed and suggest improvements. For more complex issues, technicians can work with the product designer to modify the design.
- *Get some runs on the board*: Service bureaus run printers all day, every day, and gaining some early successes with 3D printing can be highly encouraging.

Although this is a good first step, one of the key problems with this approach is that the product designer does not experience the technology first-hand, and that hands-on experience is vital in designing effectively for metal additive manufacturing. The key reasons for this have to do with the process constraints discussed previously, but also with the difficulties caused by the realities of post-processing metal prints. This is where the product designer and the engineer tend to part ways. For the engineer, the geometries are likely to be conventional and the internal microstructures are the focus; whereas for the product designer, the opportunities for innovative, complex geometries in applications are the main attraction of the technology.

Much of the literature on the topic comes from the engineering discipline and, therefore, focuses on studying the material characteristics of prints using conventional shapes. Finding design-focused literature on methodically working with unconventional geometries at an applications level in metal is much more difficult. One of the reasons is that, as an expensive process, with safety constraints, product designers generally learn through technical support. The problem with this is that the creativity that a product designer brings can, therefore, be constrained by rules developed by printing more conventional geometries designed to increase machine throughput or optimise mechanical performance.

For the product designer working on creative applications using metal 3D printing, one of the key areas of knowledge that must inform the design is about what happens after 3D printing during post-processing. Although it is a powder bed process, metal additive manufacturing has design elements that are closely related to fused filament fabrication rather than to polymer, powder bed processes as the part is connected to the base as it is being 3D printed, and this needs to inform the design decisions being made.

Product designers are more likely to be familiar with filament printing, because of the proliferation of desktop printers now in design offices rather than powder bed, so this can help. However, although there are similarities, there are also significant differences. Support structures look similar but have a different function in metal printing to filament 3D printing and so are not constructed in the same way. Their role is more to do with the dissipation of heat and combating distortion rather than supporting overhangs. In addition, in some additive manufacturing machines, the powder is spread using a heavy roller which has a significant impact on what can be printed, and how it is oriented in the build chamber. While additive technologies allow for new geometric complexity, with metal 3D printing, increased complexity also raises the risk for failure.

Post-processing complex metal geometry: A hands-on case study

When a designer can operate a metal 3D printer themselves, new insights into how to design for post-processing for that technology become clearer. In this example, the application is a customised stainless-steel handle designed for a bespoke cabinet for a public building. The cabinet was to be a high-end, value-added timber piece created by a designer-maker; thus, conventional handles would be inappropriate. To match the intent for the cabinet, metal additive manufacturing was chosen, as it allowed for the creation of a one-off or short run of a creative design, using high-end materials with a good finish. The handle was designed to meet the application requirements whilst simultaneously providing an exemplar for 3D printing for designer-makers.

Bespoke hardware is a good starting point for designer-makers specialising in timber, who may be unfamiliar with the technology and the opportunities to add value to their work. Most designer-makers buy in hardware for their furniture designs, so 3D printed hardware does not challenge their existing practice. Only a small number of designer-makers have the skills themselves, or as part of a team, to fabricate both the timber and metal components for a custom design.

The design shown in Figure 1.17, therefore, explores geometries that could not be made using conventional technology and would demonstrate the principle to designer-makers. The CAD model was built in 3D computer-aided design solid modelling software, in this case Solidworks. It could equally have been built in other similar programs, such as Fusion 360 or Inventor. The model was built by creating anchor points on a variety of planes, then adding multiple sketches on the different planes. Pathways between those sketches are built by lofting between them to create the organic shapes of the handle.

Understanding the desired print orientation for the part informed the design process. Particularly, it was influenced by the need for support material with metal additive manufacturing. Just like single-filament FFF or SLA, the supports

1.17
Handle manufactured in stainless steel using selective laser melting (Loy).

1.18
Handle showing
support material
from the selective
laser melting
process.

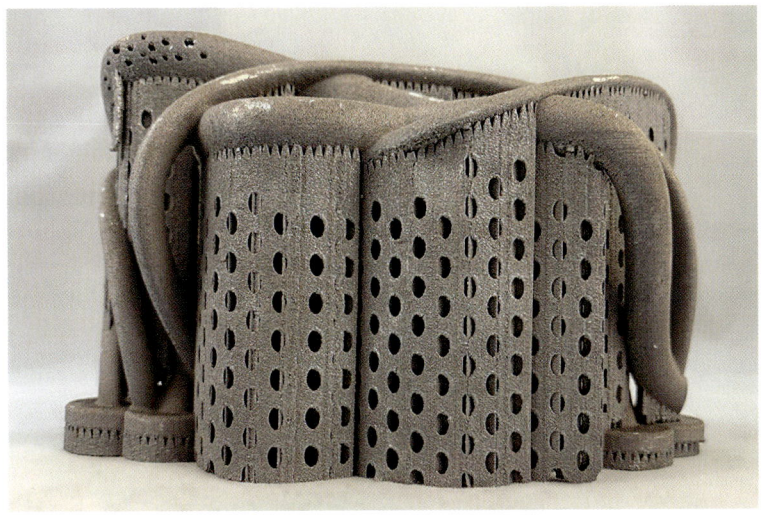

are made from the same material, and so the manual work required to remove them and clean up surfaces must be considered. This is significantly more difficult than with polymers. Therefore, the handle was designed with the intent of connecting supports to the inner surfaces, facing the drawers, so that it is not visible by the user once it is installed. The parts of the handle to be fixed to the cabinet were also designed flat, providing reliable anchor points to the build plate.

Currently, when working with service providers, the product designer will receive a single quote for the part, so the removal of support structure as shown in Figure 1.18 would not be listed as a separate item, for example. This will include the cost of the preparation of the STL, such as the slicing and packing of the print (see the points made on file preparation in Chapter 7), the volume of material used in the part, the cost of running the printer (including a portion of the powder that will have to be replaced in the next print), and the post-processing. For a part such as the handle, this last stage will be the most expensive for the service provider, despite costs rarely being broken down for a designer to see. Removal of the support structures is not as automated as the product designer might expect. The parts come out of the machine attached to the base plate (as in Figure 1.19) with some excess powder still on, which, to the greatest extent possible, is removed by vacuum and brush within the build space so it can be reused. Nevertheless, there will still be some waste, as it is still a relatively messy process.

After powder removal, metal parts must be physically cut from the build plate using machinery like a bandsaw or Wire EDM (Electrical Discharge Machining). The detail of surfaces where support material is removed require further finishing to smooth the roughness away before polishing, visible in Figure 1.19.

1.19
Multiple prints attached to the base plate (top); manual removal of support material (bottom).

For metal printing, there is frequently a considerable amount of hand finishing necessary, even for service providers. Whilst tumblers are used as a starting point for metal print post-processing, skilled workers and tools, like those present in the jewellery industry, are currently used to finish the objects. In this case, the support material was removed by hand using heavy pliers and files, often with the part clamped tightly in a workshop vice. Further finishing involved bead blasting, polishing on a linisher, and detailed polishing with hand tools. When parts like this are provided through a service bureau with a single price, it hides the issues involved in printing and finishing a product.

A greater transparency in practice by service providers would enable product designers to work more effectively with additive manufacturing technologies. In the long run, it would result in companies who invested in printers being better prepared for working with the challenges involved. It is not uncommon to speak with manufacturers who more recently purchased a metal powder bed fusion printer and were surprised by the time and expense of getting parts to look like those advertised in brochures or photographed for news and academic publications.

DESIGNER'S COMMENT

Invest time in improving your 3D modelling for surfaces and organic shapes to exploit the opportunities 3D printing provides. It doesn't matter which 3D CAD modelling software you use – they almost all have 3D printing plug-ins available now. 3D modelling is rapidly becoming the main, and easiest, method of communication for designers and engineers, and you don't want your designs to be restricted by what you can model on screen or instruct someone to model based on 2D drawings.

Additional knowledge that may not be clear for designers is that the printers themselves, from different manufacturers, have different specifications that will impact designing effectively for them. For example, the actual build space of a printer needs to be tested to ensure that the quality of the print produced is not affected by where in the build space it is being printed. This might be due to the position of a heating element, or to the performance of a particular machine. Whilst the physical print volume is easily measured, the reality is that the laser might not effectively reach the outside extremities of the build space, or that a part of the build volume needs to remain clear to allow for gas ventilation or moving components within the chamber. In addition, the powder spreader for different machines can be very different, for varying reasons. In most machines, a silicone blade is used for spreading the powder layers.

The advantage of the silicone blade is that its flexibility enables fewer delicate parts to be disturbed on each pass. However, for applications where part isotropy is essential – for example, in aerospace applications – a heavy roller spreader machine can be chosen, as it compacts the powder with each pass. The machine used in this case has an 80 kg roller. The implications of this include the addition of extra support material to ensure that in instances where the part has small islets, as in the examples shown in Figure 1.20, they are not damaged by the roller, and that the part itself is fully anchored to the base plate and will not be disturbed on each pass.

Working hands-on with test pieces in metal additive manufacturing allows the product designer to properly understand the process. This applies to both the constraints and the opportunities. As with all manufacturing technologies, the product designer needs to understand them to the point where a technician is not able to block innovation through restricting the work to conventional machine settings and past production outcomes. For the product designer, knowledge of pushing the boundaries of a technology supports innovative practice. An

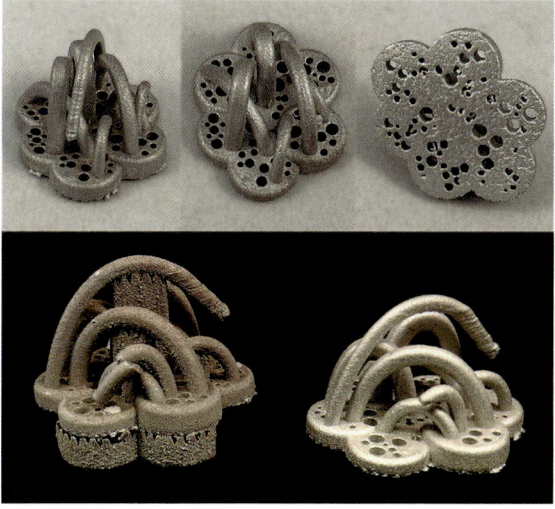

1.20
Small test pieces for understanding support material strategies for metal printing.

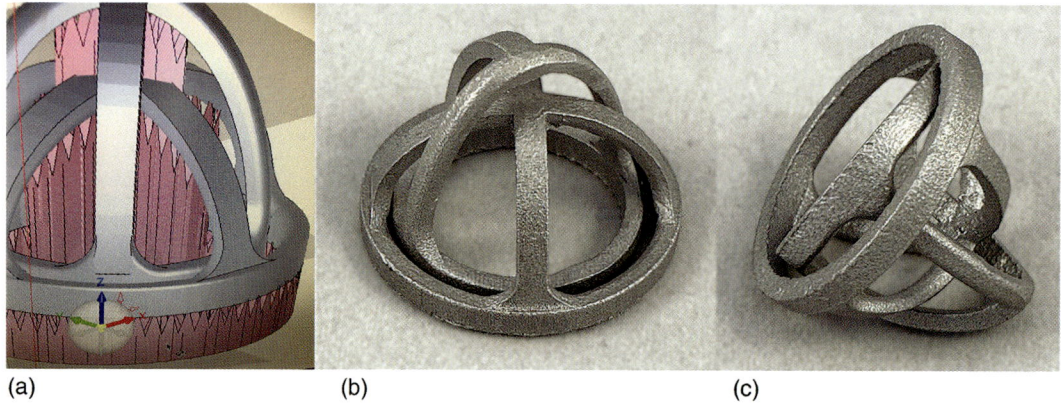

(a)   (b)   (c)

example of testing this principle is shown in Figure 1.21. Technical recommendations for working with metal additive manufacturing suggest that an assembly printed as a single part is not possible. However, with experimentation, the samples in Figure 1.21 were developed, demonstrating how building two intertwined parts on a single raft would allow for an assembly as a single part.

    Working backwards in the workflow and designing for post-processing for any of the families of 3D printing technologies is an essential element of maximising its potential for the application. Each technology has its differences, and within that technology, manufacturers create machines with individual characteristics aimed at different target markets. For the product designer, it is not enough to understand the technology in general; they need to understand working with a specific technology in context. That is, they need to understand the value and constraints of machines as they relate to a use case. Post-processing is a vital component of this professional development journey, as is the real-world understanding of each process to go from 3D CAD model to physical 3D print.

1.21
Creating an assembly with metal printing is possible and requires an understanding of support placement (a) and tolerances, informed by an understanding of post-processing (support structures removed) (b, c).

KEY POINTS

- 3D printing is not a push-button technology; product designers need to understand all stages of the process to design for it effectively.
- Test pieces help the product designer to understand working with a process and an individual machine, as there are many differences between them.
- The provided instructions for working with a technology are a starting point, but they are not sufficient in themselves for the product

designer working to maximise a technology and make appropriate recommendations to companies and clients for its use.

- As an industry, service providers need to provide greater transparency on the different stages of preparation, printing, and post-processing to educate product designers for the positive development of the industry overall.

The importance of designing for individual processes is revisited during the following chapters, with examples of practice and strategies for designing in context. Altogether, the chapters provide product designers with tools, strategies, and knowledge to approach designing for industry-level 3D printing technologies with confidence built on their disciplinary knowledge.

REFERENCES

1. (ISO). *Additive Manufacturing – General Principles – Fundamentals and Vocabulary*. Switzerland: ISO – International Organization for Standardization; 2021.
2. Iancu C, Iancu D, Stăncioiu A. From CAD Model to 3D Print via "STL" File Format. *Fiabilitate si Durabilitate – Fiability & Durability* 2010;1(5):73–80.
3. Attene M. As-Exact-as-Possible Repair of Unprintable STL Files. *Rapid Prototyping Journal* 2018;24(5):855–864.
4. Baumann F, Bugdayci H, Grunert J, Keller F, Roller D. Influence of Slicing Tools on Quality of 3D Printed Parts. *Computer-Aided Design and Applications* 2016;13(1): 14–31.

2 Working with a design for additive manufacturing (DfAM) consultancy

When product design emerged as an industrial discipline, its role was to bridge the gap between meeting market demands and creating consistency and efficiencies in production. Today, product designers still step back from production itself to view the challenges involved in designing, manufacturing, and selling a product holistically. Efficiencies are evaluated along the production and distribution supply chain, and understanding the market is as critical for product designers as understanding manufacturing constraints. This makes the role complex. Throughout their career, a product designer will need to learn new technologies and integrate them into production. However, additive manufacturing comes with additional complexities that can affect workforce development and alter market interactions and business practice, much like the development and integration of computing itself.

In entrepreneurial work, using additive manufacturing involves working with new priorities and starting points. Opportunities offered by the technology can be used as foundational principles upon which a new business operates. In contrast, working with a company where additive manufacturing is yet to be introduced, or has limited use to date, is arguably more difficult. For a product designer to effectively support the company they are working for in this situation, they need to understand the complexities involved and work to mitigate their implications. Designing a product with additive manufacturing for a company without considering these complexities can result in disruption and logistical challenges that can undermine its success. It can also be met with resistance from the workforce if not introduced properly. In practice, products designed to be manufactured using additive manufacturing for the first time in a production environment can require considerable investment both in terms of hardware and human resources.

Companies need to be aware of the implications of selecting an additive manufacturing technology, regardless of the specific technology chosen. They also need to be guided in using it appropriately, in terms of both its internal and external impacts; for example, on material sourcing and storage and on changing customer interactions. To overcome the problems that can arise, those

DOI: 10.4324/9781003122203-3

problems need to be understood and framed in a way that can be articulated, visualised, and then addressed.

The approach described in this book is based on experience of working with industry, recognising the need to see the whole situation for that company, rather than focus purely on designing a product. This "journey" can be mapped and illustrates the adoption of additive manufacturing into a business, from introducing it to a company with no existing experience of the technology, to sophisticated, radical innovations that alter business operations for companies where expertise, experience, and investment already exist. Companies will most likely adopt the technology through an intervention, such as a decision to make an investment to help future-proof a company, or through a market opportunity. Because of this, they may not plan on going through this adoption journey. However, in an ideal scenario, this "journey" provides a step-by-step approach to smoothing the way for successfully integrating the technology in production practice.

As the reality is likely to be messier, or if a product designer is asked to consult for a company, rather than build additive manufacturing expertise and acceptance from within, the first move for them would be to evaluate the company to see where it sits on the adoption map shown in Figure 2.1 and explained in detail in Chapters 3, 4, and 5.

Assessing where the company sits currently involves reviewing:

- The processes currently used, from production floor to the design office (if there is one).
- Current knowledge and expertise within the workforce, both on the factory floor and in the design office.
- The market profile.
- The supply chain and logistics.

It is also important to assess the current knowledge of different levels of management to ensure that a good understanding of what support and professional development might be needed and where there might be resistance, as well as where there might be champions across a company who can help with the transition. These champions can be at any level within a company, and ideally should include people "on the ground" who can drive bottom-up adoption rather than being purely top-down managerial decisions. Existing use of other digital fabrication technology, such as CNC routing, can help build a bridge between conventional and additive manufacturing. However, it is also important to note that a company that has invested heavily in multi-axis CNC capabilities may be reluctant to shift their investment strategy again into what could be perceived as a competing technology. Overall, these many factors will help guide positioning of a company on the journey map, ensuring that any designs and interventions are appropriate for that company.

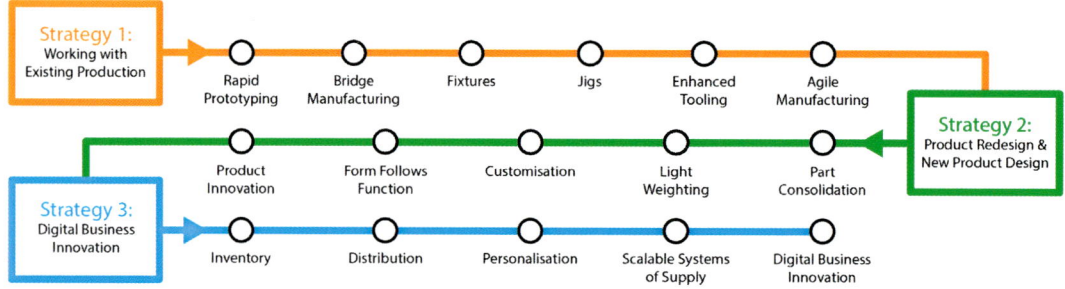

2.1
Mapping the
technology
adoption journey.

POSITIONING A COMPANY – HOW FAR ALONG THE TECHNOLOGY ADOPTION JOURNEY MAP DOES THE COMPANY SIT?

In this book, three levels of readiness are summarised in Figure 2.1 – working with existing production systems, product design, and changing business models – and discussed with prototype and product examples in Chapters 3, 4, and 5 for the introduction of additive manufacturing into a company. In response to each level, a strategy has been outlined, which is then broken down into subsections that take the company through different stages of technology adoption. These range from workforce development-centred to industry transformation. They are presented as a continuum and would be ideally followed to accumulate knowledge within a company and build support for the technology as it is integrated into practice. However, this is not essential and is highly dependent on the company, in terms of both their current practice and their short-, mid-, and long-term goals. The task for a product designer is to evaluate where the company sits on the continuum. This relates to: what skills it has in house; what experience of designing for, and working with, the technology it has, as well as the availability and suitability of the materials for additive manufacturing; and whether their market is suitable for the technology. This must then lead to identifying appropriate opportunities for that industry sector and client, then to researching relevant additive manufacturing technologies for that company and how to integrate the technology holistically into existing practice. This is a real-world problem.

The level of experience and expertise in a company needs to be evaluated to decide on where the company sits in relation to the strategic integration approaches described in this book. Whilst a company may be producing end-use parts, expertise may be limited to a small section of the workforce. To support long-term integration of the technology into production practice, and to support the adoption of Strategy 3-level digital business innovation, working through the earlier strategies with the company are recommended.

... AND A TOUCH OF IMAGINATION

The following is an introduction to the product designers at a fictional consultancy company – LND Consulting – who specialise in advising companies on the 3D printing technology adoption journey:

Droov: I'm interested in the future – the potential business impacts of additive manufacturing, what can we do with this technology? Where could it take us? How can we make the biggest impact on profits? What do we need to do to evolve working practice and business operations to get there? I keep up to date on opportunities created by digital connectivity and where to from here – exciting times!

Meeks: Design is everything to me – and I am immersed in the design world in all its forms. 3D printing opens doors for design than conventional manufacturing practices tend to shut. It brings an amazing freedom – in complexity, in personalisation, in ideas. I want to see it explored to the max. To do that, industries must know how to use it fully.

Zac: I see myself as the realist in the consultancy – more grounded. I want to make it work on the shop floor, and at a design engineering level. My concerns with additive manufacturing are around post-processing, machine maintenance, the functional viability of products. I know these things are a concern to the other designers too, but for me they are the foundations of getting any production company working with the technology properly. And any good production operation starts with its people.

The following scripts are fictional exchanges between stakeholders in a company employing product design consultants Zac, Meeks, and Droov from LND Consulting, as the organisation progresses through their 3D printing technology adoption journey.

Scene 1: *Monthly production planning meeting. Attendees include CEO, in-house product designer, production manager, and department heads.*

CEO: So, I don't know about this new technology, what's it called again?

In-house designer: Additive manufacturing, also called 3D printing. We already have some printers in the design engineering office.

Production manager: We don't need any more of those toys.

In-house designer: They aren't toys; the ones we have are desktop versions for prototyping and yes, they are low-cost versions for making models. But what we're talking about here are industrial machines.

CEO: And what will the industrial machines give us that we don't already have?

In-house designer: We can make parts more independently; we won't have to buy as much in.

Production manager: I'm in favour of anything that means we can control more of production here, in the factory.

CEO: But the reason we don't make as much on site as we used to is because it is cheaper to import.

Production manager: Not at the moment, it's not.

CEO: These are difficult times, but this will pass.

In-house designer: We should look at future – proofing the organisation in case something else happens that disrupts our supply chain. This situation will ease, but there will be others. The relocalisation of manufacturing is going to be a more important issue in the future – and we need to be more agile. We need to be able to retool more rapidly, and add value to our products.

Production manager: That sounds all a bit drastic to me.

CEO: Hmm. So, you think we need to invest in an industrial additive manufacturing machine?

Designer: Yes, but It's not as simple as just ordering one, there are lots of different machines – there are seven families of additive manufacturing technologies, and each of those has several different production methods with different machines in each case, so which one we buy depends on what we decide we want it to do.

CEO: Will we be able to add to our product range? What do the machines offer there?

In-house designer: New product opportunities, new markets. We can make products that have bigger profit margins because they are value added. We can create customised products.

Production manager: Customised? Hardly. Are we going to make each product individually? I can't see that working.

CEO: We need to understand the technology suitable for our situation – our industry, our capabilities, our market, and then we need a strategy. We need someone experienced at designing for additive manufacturing technology. Someone to help us understand how the technology could apply in our situation. A specialist consultant to advise us on what strategy to use, what additive manufacturing technology would be suitable for us, and what we would use it for, and then lead the changes we'll need to make. I'll need some convincing, though. I don't want to waste money on technology that's not right for us.

Production manager: Or that doesn't fit with our current practice – the workforce will never go for it.

In-house designer: I need to know what our competitors are doing too. Are they using additive manufacturing? I'd like to see examples of applications from competitors in our industry sector, but also, I want to see what is being done in other industries that could be applied in our market. If we can get it right, it will really help the company. We need them to understand our company, we can answer any questions that they have, we just need to know what information they'll need and who they want to talk with.

Production manager: I'll get a list of key people in the organisation they can ask advice from.

In-house designer: And they can always bring in industry experts to explain how to implement specific technologies.

The department heads agree and start suggesting key people.

CEO: And I can answer questions about our current business model and supply chain organisation. They'll need to be prepared to research our company, then we'll ask them to present us with a proposal for a strategy we can use, and we'll discuss it with the board.

Scene 2: Strategic planning meeting. Attendees include the CEO, in-house product designer, production manager, and LND Consulting. One of the LND Consulting designers, Zac, arrives.

CEO: Okay, so I've called this meeting because our design consultant wants to know more about our current production practice, and how we use 3D printing. He's recommending that we start by integrating 3D printing into our existing practice, is that right?
(Zac nods and goes to speak but is interrupted.)

In-house designer: Why are we doing that? Why can't we just add additive manufacturing machines into the production line, and extend our product range?

Production manager: Because that wouldn't work, that's why not. Our workforce isn't trained in it. The supply chain would be completely different.

In-house designer: The type of products we made would need to be different, and depending on which additive manufacturing technologies we invested in, we'd have to change our whole shopfloor – some of the machines need serious ventilation systems.

CEO: Which ones? Are those ones we need? *(Looks at the consultant, who nods and again goes to speak but is interrupted.)* Are they? I'm not ready to disrupt the whole production floor.

Production manager: Well, I'm glad to hear that.

CEO: So, you are saying there is going to be resistance from the workforce?

Production manager: I can't say I'd blame them. You are asking them to learn a whole new set of skills. And these machines are hugely expensive – over a million dollars, I've heard.

CEO: What? Seriously? This is sounding very risky.

In-house designer: Calm down. You have a transition strategy to work with existing production practices, so we can build the skills in the workforce – both on the production floor, and in the design engineering studio, don't you?

Zac: We do, and we are very aware of the sensitivities involved here, plus the workforce upskilling that's needed. This is a long-term strategy, not just a parachuted-in solution that will fail when you try to implement it.

CEO: I like the sound of a transition strategy. Okay, so we'll work through your ideas. Number one: Rapid prototyping, using 3D printing for visualisation and testing.

In-house designer: We do that already.

Production manager: No, we don't. We use those low-cost desktop printers to make visual models, but they aren't suitable for testing parts that we're going to make in production.

In-house designer: True, if we're talking performance characteristics, but we can use them to test for fit as part of assemblies, for example. Plus, our marketing department uses those models to get feedback from customers on product development, without the need to invest in moulds or other types of tooling. Much cheaper; it means we can try more risky ideas out.

CEO: Sounds interesting.

In-house designer: We could invest in a higher-quality additive manufacturing machine that is similar to the low-cost desktop ones in terms of technology, but with a higher-quality output. Then maybe we could use the parts we made on it for testing.

Production manager: No, we couldn't. A part made on one of those machines will have different performance characteristics than one made on an injection moulding machine, for instance. Just because it could perform as an end-use part doesn't mean it will have the same capabilities.

In-house designer: True, it may be better.

Production manager: I remain to be convinced on that one.

CEO: Either way, it doesn't help with the types of products we would need the prototype for, so essentially, we are agreed, at this stage there is no need to upgrade the desktop printers from visual models.

In-house designer: I don't agree. There are 3D printers that can print in multiple materials all in the same print. That would be very useful for visual models. Really step up our opportunities to get good market feedback before we tried manufacturing something new that was complex.

CEO: Interesting. *(To Zac)* What do you think?

Zac: Let's look at the options for your use case. You will need to upgrade, but again it will be part of a transition – I'm not asking you to jump in with a massively expensive investment.

CEO: So, find out what type of 3D printer can print lots of different material characteristics in the same product, and let me know what it is, how it works, and how much it costs would be useful.

In-house designer: If you can find if any of our competitors are using it with examples, that would be good, or if other industries are using it in ways you think could apply to us, I'd like to see it.

CEO: Me too. Okay, next. Iterative design development. What's that?

In-house designer: That's an interesting one. We've noticed in other companies they have started to bring out products on short runs that are then updated based on customer feedback. Once the product is how the user wants it, then the company invests in tooling.

CEO: That sounds ideal. Why can't we work like that always?

Production manager: Because mass production is based on the principle that you mass-produce – that means you invest in upfront tooling, for example, moulds for injection moulding, or a cutting tool for milling, which is very expensive, but then you can make huge numbers of parts very fast. Maybe you can make a product one at a time for a customer, but we'd be out of business pretty fast that way, as they would cost more than a customer would be willing to pay.

In-house designer: There are ways around that, but based on our design engineer's assessment, we aren't ready for that yet. Let's stick to the strategy of integrating into existing production being outlined.

CEO: Right. So iterative design, but just for use as a tool for testing the market, then production still using conventional manufacturing.

Production manager: Nothing conventional about our manufacturing, we are very innovative.

CEO: So, next suggestion. Jigs and fixtures. Anyone want to explain?

(Zac goes to speak, but the production manager jumps in.)

Production manager: Now this I am keen on. This is using additive manufacturing to support everything we are already doing on the shopfloor, just add to efficiency.

CEO: I asked Health and Safety for an opinion on this strategy, as it relates to their department. I've got them on the line *(switches phone onto speaker)*. Okay, we're ready for your review.

Health and safety *(on phone)***:** Thanks for bringing us in on this. This is very exciting technology from our point of view. Once your engineer explained to us how it could be used, we are all keen to get started.

CEO: Good to hear. More detail?

Health and safety: The key here is that the technology allows us to make products without any upfront investment in tooling. So that means we can make one-off products.

CEO: I thought we had established that would be too expensive?

Health and safety: As a replacement for production output, yes (though that's not my area), but for individual jigs, designed to help a worker on the assembly line to fit parts. From an ergonomic point of view, this is a game changer.

Production manager: And from an efficiency point of view, it makes their work faster.

Health and safety: That may save money, but really the big savings for the company are in terms of preventing injuries.

CEO: I'm happy to hear about both. Okay, thanks. *(Hangs up.)* So ergonomic jigs? Can we make those?

In-house designer: Definitely, though we'll need a bit of time to retrain the design office as we're usually focussed on commercial products, rather than the way people move on the assembly line. Plus, we may need a scanner.

Production manager: My team can help with that; we have to design the workflow and user interaction on the factory floor already. They'll all be keen on this one – no threat to jobs, just helps workers avoid injury.

CEO: *(Turns to Zac.)* I'm beginning to see how your integration strategy would play out. Get everyone on board, from the workforce to the production manager. So if we start with jigs and fixtures, which of the many additive manufacturing technologies that exist should we be using? Let me know what and why, and I'll start thinking about the budget.

Scene 3: *Planning meeting six months later. Attendees include CEO, in-house designers, production manager, and department heads.*

CEO: So, how has the integration strategy worked so far?

Production manager: It's gone very well. The new machine is reliable, the jigs are a success, everyone is on board.

In-house designer: This machine isn't a toy, then?

CEO: At that price – it may not be as expensive as some industrial machines, but it's still a significant investment, so I hope not. No problems using it then.

In-house designer: No, LND's advice to upgrade to an industrial machine using the same type of technology that we were familiar with using in the desktops made the transition straightforward. The biggest limiting factor has been the build size of the new machine – now that everyone wants to use it for jigs and fixtures, the size of the products proposed is getting bigger each day.

CEO: I'm not buying another machine already; that one set us back. Can we use it for products too?

Production manager: Same issues as before, it costs to make one off parts, how could we be commercial? I don't know what else we could use it for.

In-house designer: There is another recommendation made by our consultants, as a follow-on from the jigs and fixtures, but still a part of the integration into current production practice strategy – I think it's the next step.

Production manager: I've already told you, it's not commercially viable. The print speed is too slow; we won't make money printing short runs.

CEO: Marketing want the ability to produce short runs of products, so we can meet the needs of niche markets, as well as our usual ones.

In-house designer: Ah, but, what if we use additive manufacturing to 3D print tooling for injection moulding.

CEO: How would that give me flexibility?

In-house designer: Because we could 3D print polymer inserts for conventional manufacturing dies that are suitable for short runs, so we massively reduce the cost of tooling, making short runs viable.

Production manager: I thought you were talking about metal additive manufacturing. There wouldn't be any point using the technology for that, would there? The prints would cost as much as a die.

In-house designer: True, but according to LND, we can use the ability of the technology to print moulds that have conformal cooling channels built in.

CEO: What would that do?

Production manager: Speed up the process – cooling time would be reduced, increasing efficiency. You know, that's a good idea.

CEO: I don't mind looking at the machine we've just bought for moulds for short-run prints, with conformal cooling channels – is that possible?

In-house designer: Yes, we have a dual-filament machine with a soluble support structure, so channels through the part would be possible.

CEO: Not keen yet to buy a metal printer. I've heard they cost nearly a million dollars, or even more.

Production manager: I'd need to know more details before we even thought about that. What is the build size of these printers? What metals can be printed in them? Can you print more than one metal in the same build? How expensive are they to run?

CEO: I'll ask LND to find out some more details on metal printing, just so we know what the implications would be. In the meantime, they'll need to help the in-house design department to understand conformal cooling better as part of designing short-run moulds.

In-house designer: And I'd like to know who is using this approach already, so let's ask for examples.

Scene 4: *Planning meeting six months later. Attendees include CEO, in-house designers, production manager, and Zac from LND Consulting.*

CEO: Welcome back! I hear your work as a DfAM consultancy has been going well; quite the reputation you are building there.

Zac: Good to be back. Nice to see everyone in the factory too.

Production manager: Don't forget we were part of building that reputation in additive manufacturing in this area.

Zac: We won't; it's been great to see the work you've been doing.

CEO: Yes, well, we are glad you are back again to see. In summary, we have been using high-end, fused filament fabrication for the last year in support of existing production practices, and we're happy with the results.

Production manager: We have ergonomic jigs throughout the assembly line.

In-house designer: We produce short runs based on 3D printed polymer inserts.

Production manager: And we have even outsourced fabricating metal tooling for small moulds with conformal channels that we run cold water through – made a significant difference not only to the production speeds we have, but also on the quality of the parts. Being able to control the temperature within the die has really gotten the workers into the technology. We're getting a lot of interest. It's making the workforce keen to look at getting metal 3D printing in-house.

In-house designer: Now we want to go to the next level with the technology. We want to go beyond integrating it into existing production practice, and instead redesign parts to be actually 3D printed. I understood your reluctance to recommend that when we started, but I think we're ready now. We have a workforce that is enthusiastic about the technology and understands the basic principles. Now we need your help with product redesign as a strategy.

Zac: I can see that, it's good that everyone is enthusiastic – and confident in the technology.

CEO: So, what do you suggest? How should we start this next phase? Can we take our existing products and 3D print them?

Production manager: I told you, that makes no sense.

In-house designer: Then how do we use the technology work for us for end-use products?

Zac: I agree, you are ready for Strategy 2, the next phase of technology adoption, but let's make sure we are doing it step-by-step and that we can help you at each stage of the process. We will start with product redesign, looking at improvement potential for existing products, then move onto product innovation.

CEO: I need LND to map out a strategy for product redesign, and I need to see examples of how our competitors are using it in commercial applications.

In-house designer: Plus, I need to know more about software, what we need going forward, and how we use it.

Zac: We can advise you, find what's right for you – don't forget to budget for additional software as well as hardware.

CEO: So, based on our specific target market and organisation, I'd like your recommendations and justification for each technology you are thinking about and your estimates for the cost of each stage of integration. The first phase of integrating the technology went well, but this stage requires much more investment, so it needs to be based on your research, and I need to see where the gains might be – financial or otherwise.

Production manager: And what the problems will be too.

CEO: That's true. If it is not appropriate for us to move into this phase of adopting the technology, I'd rather know now.

Zac: That's what we're here for. Trust the process, and, if you are ready to make a significant investment and think about making a major adjustment – because that's what it could involve – you'll be building value-added products before the end of the year. Don't worry, though; we'll take it step-by-step, and you can decide it's far enough at any point.

Scene 5: *Follow-up meeting four months later. Attendees include CEO, in-house designers, production manager, and Meeks from LND Consulting.*

Production manager: We had to call you back in to help us to settle an issue that we are having with your product redesign strategy.

In-house designer: It's a difference in approach, that's all. Essentially, we need to better understand the use of topology optimisation as part of design for additive manufacturing.

Production manager: It isn't part of design for additive manufacturing, it defines design for additive manufacturing.

In-house designer: No, it doesn't. It's one part, sure, but it informs product redesign; it's not all there is to it.

Production manager: This makes no sense to me. Surely topology optimisation should drive design, not just inform it.

In-house designer: Absolutely not. Finite element analysis allows us to understand the part, and with topology optimisation we can reduce the weight whilst retaining material where we need it to keep the performance characteristics required. But design for additive manufacturing must consider designing for whichever additive manufacturing technology is being used, and also designing for the application, which may be influenced by other factors.

Meeks: You are both right for your particular part of the process. For production considerations, I can see that topology optimisation is fundamental, but it is also the case that for design, it is just one element of the process – it informs but doesn't determine the outcome.

Production manager: Hmm. Well then, I need some clear guidelines and examples. For our industry situation, I need you to give me a summary of key design for additive manufacturing points that I can give to the workforce – I'll need examples to show them too.

Meeks: Absolutely. I can provide those for you and your team.

In-house designer: These have to relate to the specific technology you have recommended for the product redesigns and take into account issues such as build orientation. We need to be able to agree on these things.
(CEO arrives.)

CEO: Sorry I'm late. Where are we up to? Have we discussed value adding yet? I am keen that we don't just redesign our existing products for additive manufacture, but also look at new opportunities the technology can offer.

Production manager: Such as?

In-house designer: I was thinking maybe more complex geometries, not possible with existing processes – there could be ways of value adding through that. Or personalised products?

Production manager: What is the difference between customisation and personalised products?

CEO: Customisation is where we create the parameters for a product that can be adjusted by the user to meet their individual needs. Personalisation is where the product is created specifically for them.

Production manager: I'd have to see where that's being used to see how those would work differently. Also, I still don't see how they can be financially viable.

In-house designer: Can you show us some examples and explain how it might work?

Meeks: It's great that you are keen! But I would urge you to slow down a little. Customisation and personalisation involve quite different ways of working, and my advice is building your knowledge and experience in this stage of technology adoption before looking at the next stage, as that involves changing business operations. There are many opportunities for working with this stage that we can still explore, and it's good to build that in-house expertise during this phase before we look at taking things any further.

CEO: Sounds good to me.

Scene 6: Designer Droov from LND Consulting presents an update on where the company is and where it could go next in its additive manufacturing technology adoption strategy.

CEO: I don't know about this report. It makes me very nervous.

Droov: I can understand that it is very different.

In-house designer: I think the ideas are exciting! It's a whole new way of working!

CEO: That's what I'm worried about. This is not just about manufacturing; this strategy is about changing business models. I don't know if we're ready for that. It is far more high-risk than the other strategies.

In-house designer: A paradigm shift in manufacturing.

Production manager: You don't need to say that as if it is a good thing. Some of us have spent our working lives building expertise in traditional production practices. I don't think I'll be the only one in the company with misgivings.

CEO: It is radical, no doubt about that. But is it feasible, and would it put us ahead of the pack?

Droov: It's feasible if you plan it properly. It does need careful consideration.

In-house designer: There are elements that are less high risk than others. The inventory idea, for instance. That's straightforward.

Production manager: Replacing stock with digital files? That worries me.

In-house designer: It's not all stock, just spare parts for products that we no longer make. We don't then have to store them for seven years as we do now.

CEO: That would save warehousing costs.

Production manager: It would save on the money we spend on dumping those parts at the end of the compulsory period.

In-house designer: And we could send the digital file to wherever that part is needed, saving distribution costs.

Production manager: I don't like that as much. How can we guarantee the quality of the part if it's not manufactured here?

CEO: We already buy in parts other people make; we get quality assurances there. What's the difference?

Droov: There are different quality control measures that need to be put in place, and we'd have to work through them carefully, and bring your suppliers into the loop.

CEO: Okay, so that's one aspect to look at. What about the others on this list?

Production manager: Well, I'd like to understand the idea of changing the customer interaction. What does that mean?

In-house designer: It means we start with the user needs and interaction and build a new business model based on that.

Production manager: That just does not sound realistic.

CEO: Did you see the GE Additive manufacturing example in the report? The biggest disruption was in the supply chain. That turboprop engine they did reduced the number of parts from 288 to 12, and the companies in the supply chain by some big number. Even the engineering team lost jobs. You don't want to lose the workforce.

In-house designer: But this could open a whole new raft of commercial opportunities, with high-value products that are light on resources – and that would be where the workers would be focused, and we would just be freeing up those resources, not losing them. Smaller supply chains would be good for meeting our sustainability targets. And we could create customers for life with products that they are more invested in, that can adapt as their needs change.

Production manager: You've lost me there.

Droov: That's the shift from customisation to personalisation.

CEO: Let's keep this real. What are the implications of rethinking business practice for additive manufacturing at this stage? We need more detail. We need examples and facts that we can base decisions on. At the moment, it all looks very interesting, but also a bit speculative.

In-house designer: Speculative is what we need right now – and what we're ready for. We need to rethink our business in the current context to be competitive going into the future. Incremental change isn't good enough. We need to look at the circular economy, design for disassembly, responsible material choice, reducing our carbon footprint, being more agile and able to pivot to where the demand is, relocalisation of part production, value adding, designing for additive manufacturing rather than just printing existing designs …

CEO: Alright, let's not get ahead of ourselves. I agree we need to be imaginative, innovative, but it's not that easy to think differently about current practice.

In-house designer: Then let's imagine where we would like to be in 10 or 20 years' time, then work backwards. Could you do that? Imagine a future where additive manufacturing had taken over from conventional production? What would that look like?

Droov: We could run some scenarios through with you, do a bit of back casting, where we look at where you could be, and work backwards to create a timeline on how to get there. The strategy is in stages, just like before, and when we get to the more radical rethinking of practice, we'll do the work, go through planning activities, and make sure you are ready. It's time for change …

NEXT STEPS – THE TECHNOLOGY ADOPTION JOURNEY

The following three chapters provide detail on the different stages a single company could theoretically go through if the technology adoption pathway was worked through step-by-step. Realistically, this is rarely – if ever – going to be the case. Nevertheless, the product designer can use the framework as a communication and organisational tool to work with a company to understand where on the adoption journey they are, and where the industry sector they operate in may be. Critically, it also allows the designer and the company to differentiate between the stages of technology adoption to support their integration of 3D printing more effectively into practice. The framework outlines a step-by-step approach to integrating additive manufacturing into production practice, from the least disruptive to the most innovative ways of working. Cumulatively, they provide the product designer with a portfolio of practice in design *for* additive manufacturing (DfAM) and designing *with* additive manufacturing (DwAM).

At the start of each strategy chapter, there is an introduction to a fictional company that is at that point of the journey mapping for 3D printing technology adoption. The intent of this is to help visualise placing companies on the mapping diagram and use that as the basis for recommending which of the three strategies to work with. At the end of each of those three chapters (3, 4, and 5), there is a hypothetical response to the key recommendations from a product designer and what the company decided to do.

3 Strategy 1

Working with existing production

Strategy 1, summarised in Figure 3.1, starts with the most easily accepted approach to using 3D printing in industry at this time and works through successive approaches to using the technology, building expertise, and sophistication in the applications. The intent, however, is to work with existing production practice only in this initial stage, to enhance practice without causing the disruption or resentment that can occur when new skills are required by the workforce. Using the technology for visual and physical prototyping, for example, to help with decision-making for existing production practices avoids many of the more difficult considerations needed when detailing for 3D printing at this stage. However, not all new detailing can be avoided, particularly when using the technology for the potentially challenging applications included in this strategy, such as bridge manufacturing or tooling.

Designing effectively for the appropriate additive manufacturing technology and product use will affect whether the results are successful or disappointing. Getting this stage right can make a difference regarding how the technology is perceived further down the track when subsequent stages commence. In many ways, getting this stage right is also critical to the journey of a company into subsequent stages, explained as Strategy 2 and Strategy 3, in Chapters 4 and 5.

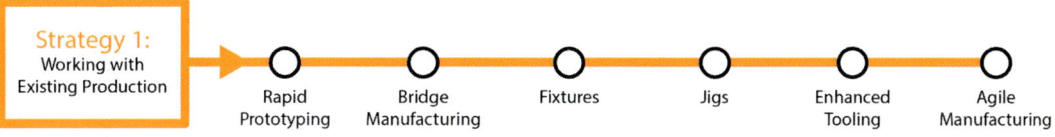

3.1
Overview of
Strategy 1
approaches to 3D
printing.

THE FAMILY FIRM: A COMPANY NEW TO ADDITIVE MANUFACTURING

Madoc Manufacturing is a family firm, run by Harry Madoc after taking over from his father, Lewis (who still refuses to retire). Harry, his wife Sara, and his daughter Erin all work in the company, and they employ 18 other workers in different production aspects of the company. The company operates on the same

DOI: 10.4324/9781003122203-4

3.2

principles as those established by Lewis. Harry and Sara contend that pride in the product, quality, and service are the foundation for their success. Sara joined the company after she and Harry were married and insisted on working at each stage of production to learn the business properly, and so too did their daughter during her own training. Madoc Manufacturing is an attractive workplace, where Harry and Sara treat employees fairly and are very committed to training them to a high standard. As a result, the skilled workforce is in demand elsewhere, but they stay for the opportunity to work hands-on during critical parts of the process rather than operate automatic machines (Figure 3.2).

Erin attends university on a part-time basis, studying engineering. She has excellent 3D modelling skills and has become very interested in the possibilities that 3D printing could provide for the business. Harry, Sara, and Lewis discuss the potential at length but are undecided how, or whether, to proceed. Lewis is particularly worried that the whole basis for the company, the skilled workers they employ, will be undermined by what he deems a "pushbutton" technology. Harry is thinking about the bottom line and whether the technology could save them money, or whether it will be a costly mistake. Simultaneously, Sara, having read as much as she can about the technology, is worried about the overwhelming number of additive manufacturing technologies and the possibility they may choose the wrong one.

Erin and Harry hold an informal meeting with the workforce during a tea break and they decide they need advice on not only what could be made with the technology but also how it would fit with the current operations. They also want to know how it would bring people along with its adoption, rather than alienate anyone. They report back to the family that this is a tricky situation, and it would be worth investing in advice from an external design consultancy with specialist skills rather than relying on their own research. The family agrees, and they contact LND Consulting, a young consultancy specialising in product design for additive manufacturing. They book a meeting. Zac from LND Consulting visits the workplace, takes time to meet the workforce, see the products

being made, and understand what techniques are used in production. He also
finds out about their market, supply chain, and distribution mechanisms. The
family explain their ambitions as well as concerns over the technology. It is clear
they are conflicted about how to proceed. After considerable discussion, Zac
recommends he works with the company using Strategy 1 and explains his
reasoning.

WHY WORK WITH STRATEGY 1?

If Madoc Manufacturing were a start-up company, the use of additive manufacturing could provide an opportunity to build in agile manufacturing right
from the start to make a small company nimble and, therefore, competitive.
Further, it would enable it to respond to short-lived market changes through
increased ability to produce short-run, or bespoke, niche products. For an
established manufacturer like Madoc Manufacturing, however, where conventional manufacturing has been in place for a considerable length of time
and skilled workers contribute to the business' reputation for quality products, the introduction of additive manufacturing can be problematic. New
technology is rarely seen as beneficial in terms of improving on existing practice because it is largely associated with far more disruptive practice and
expectation.

With innovations based on established technology, the accepted ways of
working are not being challenged, just supplemented. On the other hand, additive manufacturing can frequently require a very different way of thinking,
designing, and making. It requires new skills and can result in new ways of
interacting with customers. Therefore, when considering recommending a
strategy for the introduction of additive manufacturing to any company, it is
important to understand what level of adoption the company is ready for at that
time. The three stages of technology adoption discussed in the next three chapters cover working with a company that:

1. Is new to the technology – Strategy 1.
2. Has a foundation of knowledge and experience, ready to take the next
   steps into product redesign – Strategy 2.
3. Or, that has considerable expertise in working with the technology and is
   looking to maximise the opportunities it provides using innovative business
   practices and new products – Strategy 3.

Thus, Strategy 1 (working with a company new to the technology), enables the
introduction of a progressive, cumulative approach to the integration of additive
manufacturing into a company that has had little or no involvement with the technology to date. The intent is to support an enthusiastic adoption of 3D printing into
an established traditional manufacturing facility – a complicated task, as there
can be considerable reluctance or even resistance to additive manufacturing. This

resistance may stem from observations that additive manufacturing is not straightforward, is multifaceted, and covers a wide range of technologies, each with their own constraints and opportunities. Further, there is a steep learning curve which requires new materials to be sourced, and in obscure quantities at first. Simultaneously, detailing will be very different and not instinctive to someone who has worked in conventional manufacturing for some time, and so too the software and workflow, which can undermine the willingness of personnel to take it on. Anyone with accumulated knowledge in a particular field is likely to find the transition to this technology unattractive.

For Madoc Manufacturing, this first strategy then becomes the most appropriate, and important. Tempting though it is to focus on designing new products immediately, a move frequently encouraged by innovators in management keen to make the move to the technology, resisting that pressure and the temptation of presenting a "resolved" product can lead to longer-term innovation that is sustainable, over a one-off flurry of activity. Bringing everyone along on the journey is more likely to succeed over time, but it will be slower, which should be understood and acknowledged by all involved. The "journey map" presented in this book can help with this conversation.

Each stage of Strategy 1 is designed to engage and influence the attitudes of a manufacturing workforce not very familiar with 3D printing. They also enhance the accumulation of skills and develop a sense of ownership of the technology within the organisation. It allows the product designer and the company to evaluate the level of expertise in the company, and to identify any concerns of the workforce in relation to working with additive manufacturing. This strategy also enables preparation for the introduction of the technology through a range of positive uses, which, it is hoped, will be perceived as non-threatening to anyone with those concerns. The following sections discuss these different stages more specifically.

The steps outlined in Strategy 1 are aimed at engaging a company like Madoc Manufacturing in using additive manufacturing from a standing start. The recommendation is to begin by using it to build test prints for products to be made using conventional manufacturing, rather than for new product ideas. The second step is to use the technology to test the market to iteratively improve a product – but still a product ultimately to be made for mass production using conventional production processes. The next step is more significant, as it involves the company's ability to start designing specifically for additive manufacturing, but only to support existing production practices. Erin's engineering and 3D modelling skills will be invaluable at this point, as an effective way to do this is to introduce bespoke jigs and fixtures to the production workflow using additive manufacturing. This is a way to use the technology for products new to the company, and to build the skills needed to do so, but still only in support of existing practices, so it is not seen as a direct challenge for anyone whilst the workforce builds familiarity, skills, and acceptance. This approach should also result in personal benefit for individuals if the new jigs and fixtures prove to be effective – and introduce the concept of personalisation made possible with additive manufacturing, relevant in Chapter 5, where it forms the basis for commercial products.

Once the workforce at Madoc Manufacturing is more confident, and hopefully enthusiastic, about using the technology in these ways, the next stage recommended in Strategy 1 is to see where it could be possible to enhance tooling using additive manufacturing for added value – again, to existing production practices. This might involve adding conformal cooling to a tool, for example, providing quantifiable improvements to production through reducing the time it takes for a part to cool down, or managing the change in temperature across the part. It could also involve the introduction of agile or flexible tooling, such as using 3D printing inserts in tools, or producing short-run, stand-alone 3D printed tooling to supplement practice.

Depending on the manufacturer and the context, product designers may lead the adoption of all these approaches incrementally into a company, or choose to implement one approach that illustrates the potential of the technology to the company and build buy-in and expertise prior to moving on to Strategy 2, which focusses on product redesign and innovation. For Madoc Manufacturing, working through Strategy 1 would align with their values, as providing bespoke jigs for workers and improving the performance of tooling are both ways of differentiating the quality of their production practice. Even Lewis was cautiously interested.

DESIGNER'S COMMENT

Working with a company new to 3D printing involves helping workers on the production line to build confidence in the technology as well as a store of new knowledge on how to use it effectively. Rushing in with new product ideas before all stakeholders are on board, with the right skills, can result in tech adoption "crash and burn" before it has a chance to be properly understood.

RAPID PROTOTYPING

3D printed prototypes played a vital role in the development of the MagnaLatch Series 3 pool safety gate latch, manufactured by D&D Technologies and designed by Australian consultancy Ivoke Pty Ltd. All the white pieces in the exploded view were 3D printed by a service bureau using SLS, with the nylon material having similar properties to the injection-moulded, glass-filled nylon used to manufacture the final design.

This allowed for prototypes to be used for real-world testing – for example, development of a gate slamming rig to ensure the design would be robust and able to meet numerous international standards related to adjustability and gate fitting requirements. SLM of stainless steel was also used to prototype some of the custom metal components, such as the worm drive that allows the latch to be adjusted horizontally.

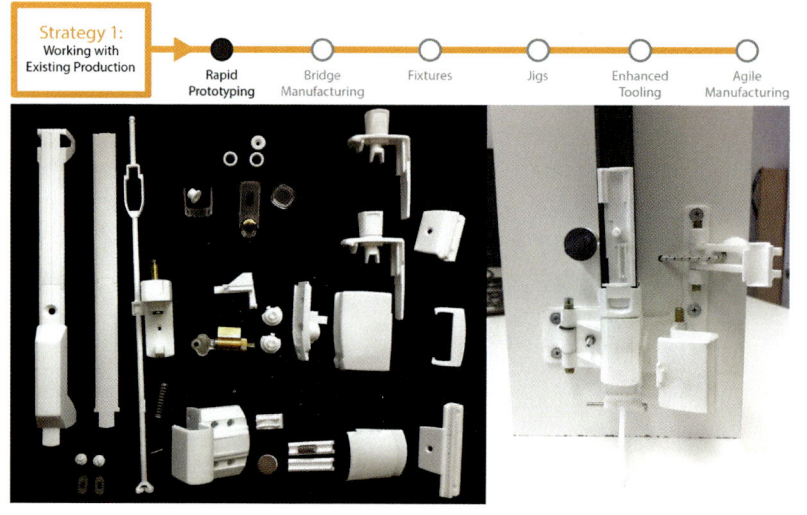

3.3
Example of rapid prototyping – MagnaLatch Series 3 pool safety gate latch. (Image courtesy of Glenn Smith, Ivoke Pty Ltd.)

For a product designer new to using 3D printing, whether as an undergraduate during study or as a practitioner with expertise in the myriad of other technologies that form the body of knowledge on manufacturing, rapid prototyping is a good place to start (Figure 3.3). *Rapid prototyping* was, in fact, the original name for 3D printing based on the limited applications possible with early hardware and materials, finding most use as a prototyping tool [1]. Despite the proliferation of the technology today, the most obvious starting points may not necessarily be the best choice, depending on the circumstances.

Most people using a 3D printer for the first time will start with the most accessible printer – in other words, the cheapest, most understandable, and least dangerous printer to use, which is the desktop filament printer, also known as fused filament fabrication (FFF) or fused deposition modelling (FDM). The starting price is very low, usually in the hundreds instead of thousands of dollars. In operation, the filament is heated and extruded, much like a glue gun, and the part builds in layers, beginning on a build plate. Nevertheless, as introduced in Chapter 1, a single-filament desktop printer is arguably one of the hardest technologies to learn to design successfully for and has more in common with metal powder bed printing than other, less expensive additive manufacturing technologies. This is because designing for the single-filament printer involves working with support structures, understanding the impact of changing the settings available, and working with build orientation.

When working with 3D printing, the idea that it is possible to print just about any geometry does not readily apply to single-filament extrusion printing. This is particularly pertinent when considering the role of prototyping, which is to test an idea, gain feedback, iterate, and move closer to a final product that will ultimately be manufactured using a different process. Therefore, the design may naturally include details and features that are not well suited to 3D printing, for example, fine text or logos embossed on a surface, or large flat surfaces. In this case, the most flexible 3D print technologies (e.g., selective laser sintering or multi jet fusion) which provide good surface finishes and accuracy perform best. If the product designer has access to a service provider, either online or locally, and has the time to wait for the print, then specifying for a proven industrial technology such as this is more likely to produce reassuring results that do not deter a company from further engagement with the technology [2].

Over time, however, having a quality desktop printer on hand and building a working knowledge through hands-on experience of its use are valuable and highly recommended. This is not only because the desktop 3D printer becomes a useful studio aid to design and development practice, but also because, ironically, it is one of the most difficult 3D printing technologies to design effectively for, as it requires a good understanding of how to design to avoid support structures in the print. Therefore, if workers learn how to produce viable models using a single-filament desktop printer, they are likely to then find subsequent forms of additive manufacturing much easier to work with. Having such a printer on hand can also increase the speed of product development, as it eliminates the need for waiting for external service bureaus to deliver parts before being able to progress a design further.

While costs of SLS printers, for example, continue to decline, and mean that desktop versions are already comparable to commercial-quality FFF machines, the additional complexities of post-processing SLS parts as well as maintaining the machine mean the FFF is likely to remain the dominant in-house 3D printer for some time to come. The key is to introduce this at an appropriate time because quite often, FFF parts can be underwhelming for people working regularly with traditional manufacturing technologies or craft-based techniques. Powder bed fusion can help bridge this gap for rapid prototyping.

BRIDGE MANUFACTURING: ITERATIVE DESIGN OF END-USE PRODUCTS FOR SHORT-RUN, MARKET TESTING

Although the product shown in Figure 3.4 was designed to be made long-term using conventional manufacturing, the first few examples were produced using 3D printing. This was so that they could be field tested, and any adjustments could be made before investing in tooling. This meant it had to withstand everyday use and, therefore, was more challenging to produce using 3D printing, as it had to behave as an end-use product without specifically being designed for the technology. As a result, the product was fabricated using SLS. This allows for the most realistic evaluation of parts prior to full production.

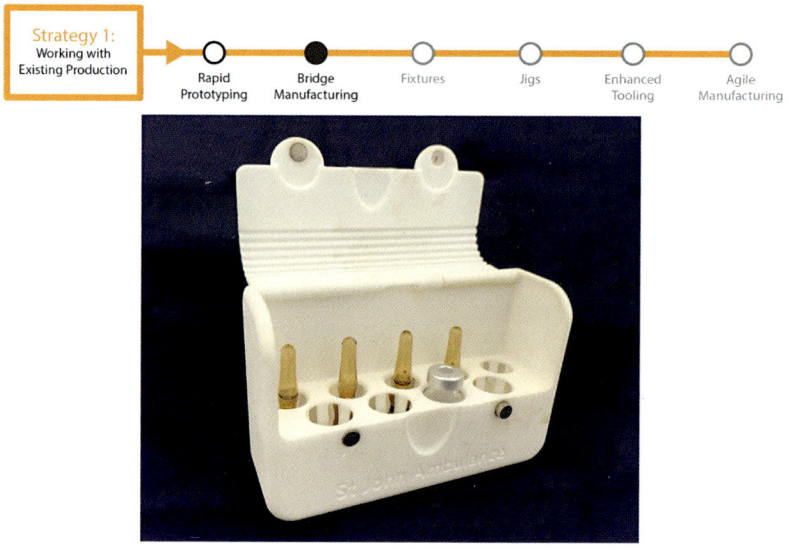

3.4
Example of bridge manufacturing: drug delivery storage container (Diegel).

For the last century, product designers have had to work within the design limitations imposed by the cost of tooling. A prototype can provide only a certain level of insight on the response of the market to a product. However, the cost of tooling can be very high, making commitment to a final design a high risk. Using bridge manufacturing, the final design can be tested on the market prior to that commitment. Although this can be a high-cost option in the short term and not appropriate for all products, as the product still must be functional and the design is not adapted to 3D printing, it can mean avoiding a costly mistake once tooling has been commissioned.

This stage of Strategy 1 – bridge manufacturing – may not be appropriate for every company, but where it can be used, it is a useful step in introducing the shift in thinking from mass production to mass customisation, concepts explored later in discussions on Strategy 2 and Strategy 3. In bridge manufacturing, an end-use, short-run product is produced to test the market prior to investing in tooling for mass production. For this to work, the materials and design must be compatible to both the conventional production method and additive manufacturing. Even if only a few specific parts are trialled in this way, the long-term benefit of gaining market feedback before investing in tooling is considerable.

This approach illustrates a process change in the relationship between producers and consumers. This is largely owing to digital technologies, even where the final product sent to market is a mass-produced item manufactured using conventional techniques. Essentially, it allows the designer and manufacturer to test a product on the market and tweak it as needed prior to launching the final version. This facility not only mitigates against disastrous details that undermine the viability of the product on the market, but also allows the designer to push the boundaries of the design to some extent. It provides a feedback mechanism for designers and manufacturers during the final stages of product development, and a level of confidence on the response the product will receive.

As the product concerned needs to be able to be reproduced using additive manufacturing whilst still destined for mass production, there may be limitations in its applications. Further, the model must be printed using viable materials rather than purely prototyping materials, which may not always be possible. However, where it is possible, it can provide valuable feedback on the product before it is launched. Bridge manufacturing also accommodates the production of variations in response to market feedback before selecting a product to commit to. A company can use this feedback to be more confident of its decisions. This is different to rapid prototyping in that the product, or variations of the product, exist as finished designs intended to look and feel like mass-manufactured goods ready to be sent to production. Rapid prototypes may look and feel like finished designs but typically lack manufacturing detail and may not have the mechanical qualities of a final product.

By using this within Strategy 1 at this early stage of a company's technology adoption "journey," the approach of bridge manufacturing described here is limited to a checking process for small detailing for conventional production rather than as a dynamic prototyping tool for design development (discussed in Strategy 2) or a new way of doing business (discussed in Strategy 3). Because the aim of Strategy 1 is to incrementally integrate the technology into existing practice in a positive way, the intent here is to build knowledge and confidence whilst still re-enforcing established production practices. It is, therefore, more about workforce development than product design.

This approach is not widely used in conventional manufacturing facilities at this time, but it could provide a useful bridge between viewing additive manufacturing as purely a prototyping process and viewing it as viable for end-use parts because it is applicable to existing production practices and is needed only for short test runs. This is increasingly possible as additive manufacturing materials and processes better replicate what can be achieved through traditional fabrication. For the product designer working to integrate the technology into a business, this can be a method of introducing the technology by stealth whilst building confidence in the process through its ability to provide early-stage feedback to the manufacturer.

For a product designer working with this approach, it may be that the part selected will be over-engineered for its application. For example, it may be produced in metal when the final part would be in a polymer, or it may be that wall thicknesses have to be adjusted to compensate for the change in process and material. These differences would have to be tested. While it may not seem a viable option in isolation, as an exercise in workforce development, it can be a non-threatening way of working through designing for the technology without having to look at designing parts that exploit the characteristics of the technology for a viable business case to be made.

Although suggested here as an integration strategy for SMEs and large-scale manufacturers looking to introduce additive manufacturing into their workflow, examples of this approach are seen mostly in small products by entrepreneurs at this time. Online forums and blogs provide discussions on products launched online that evolve based on feedback from customers. Crowdfunding websites have played a part in this as well, with bridge manufacturing used to get products to market quickly and gain feedback or deliver a successful campaign to a small group of backers before shifting to mass production. Because of this, there is a blurred line between bridge manufacturing and iterative design, as discussed in later chapters.

Working with bridge manufacturing requires the identification of parts that are suitable to be both conventionally mass-produced and produced as a limited-run item using additive manufacturing. As this strategy here is working with existing production, it may be that the part is outsourced to start with (e.g., service bureau), as it is not presented as a viable business proposition to invest in industrial additive manufacturing technology for a very limited number of parts. However, if a company is investing in an industrial process, it may be a way to improve workers' understanding of the technology without disrupting the established production line. It will help shift the perception of the technology as one for prototyping only, to one that can produce end-use parts, however expensive those early forays may be. These changes in understanding are arguably far more valuable long-term to a company than the development feedback they receive on the product itself.

DESIGNER'S COMMENT

The #1 rule of 3D printing – should you be 3D printing in the first place? When developing an idea, always use the simplest, quickest, and most effective way of testing your idea. And 3D printing is not always the answer. Laser cutters, scissors, cardboard, wire, paper, and modelling clay are still the key tools for early-stage concept work.

FIXTURES

An example of using 3D printing for developing bespoke fixtures is shown in Figure 3.5. These welding fixtures were designed for an agricultural company that was new to additive manufacturing. The design engineers, now young graduates, recommended that the company follow a Strategy 1 approach, creating products to support existing production practice whilst building familiarity with the technology in the company. These fixtures were designed for 3D printing to the extent that they needed to be, but as far as possible, they were based on familiar

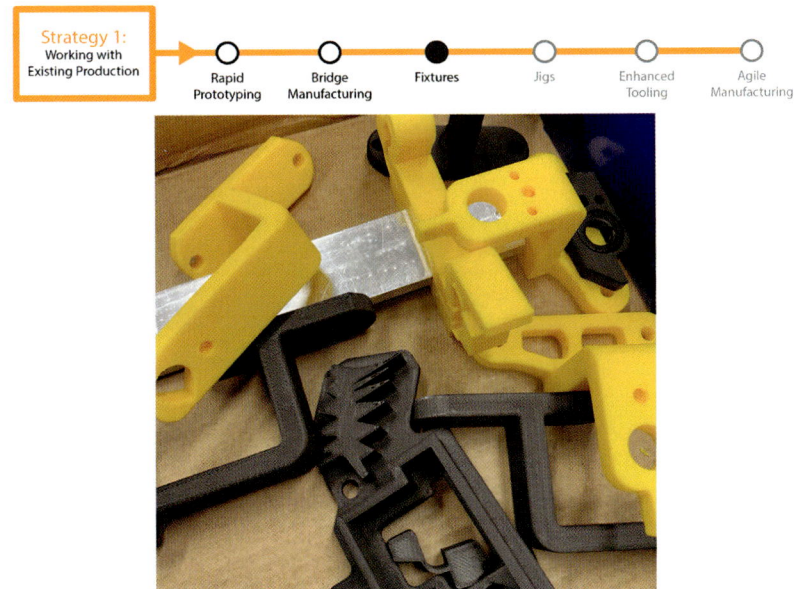

3.5
Example of fixtures for welding (by Jordan Richie and Daniel Rolfe).

forms to build confidence among workers in the company. Design considerations included:

- *Build orientations to maximise performance.* The prototypes were developed using low-cost desktop printing and PLA. The final products (shown in black) were built on a Markforged FFF machine using chopped carbon fibre embedded in the material extrusion.
- *Responding to testing.* This included resistance to high temperatures and design details to reduce stress on corners.

Fixtures are used to hold a workpiece in place during manufacture, and therefore, this hardware is a good selection to work with in cases where the engineering capabilities of additive manufacturing are a point of concern within the workforce. Fixtures allow for engineering design specific to the task to be explored which can provide a "safe" basis for testing and understanding the capabilities and limitations of the technology. Fixture designs do not need to be sophisticated – although they can be – so they can provide a good starting point for employees to learn 3D computer modelling skills at the same time as learning about 3D printing. These skills will be useful for subsequent stages of the technology adoption journey for a company where design for additive manufacturing rules must be applied.

Any manufacturer with assembly line production could explore the potential of custom-made fixtures using additive manufacturing. For those with parts that are adjusted to clients' needs, additive manufacturing is particularly useful, as the requirements of fixtures in this situation frequently need to be customised to a particular job and are, therefore, subject to change. This use is a very good starting point for exploring the capabilities of the technology under the localised conditions for the fixture. It also has the potential to be used as the basis for developing a more parametric modelling approach to the use of additive manufacturing, which builds to the customisation opportunities within Strategy 2. Parametric 3D computer models are constrained by relationships between features rather than relying completely on dimensions [3]. Critical dimensions can be specified, while some can be adjusted to fit the individual job and later altered without affecting the integrity of the part.

If a fixture is needed on a regular basis that shares some commonalities, but also needs dimensions adjusting to fit, then this provides the starting point for a company to build a supply system for itself where a worker can specify a print based on need, without starting from scratch each time. To do this, a company can start to build an interactive library of parts based on parametric modelling where there are fixed dimensions and those that are adjustable. Rather than undergoing engineering testing individually, the parts could be physically tested at the extreme ends of the use of the model. This provides the user with a range of approved dimensions on a sliding scale rather than incremental changes to the dimensions of the part. A good parametric model will allow a user to modify specific dimensions within an approved range based on testing and experience rather than adjusting dimensions freely and potentially creating fixture models that cannot be printed or are likely to break. For the product designer, this is a long-term approach to introducing a company to adoption of 3D printed fixtures. It provides a good foundation to future practice working with the technology.

Another reason that fixtures are an interesting starting point for a manufacturer is that they provide a good opportunity for moving the company from working with low-cost, studio-based, single- or dual-filament desktops to semi-industrial machines, such as carbon-fibre, dual-filament 3D printers. In terms of cost, these sit partway between high-end desktop printers and low-end industrial ones. They use chopped-up carbon fibre in a substrate (e.g., nylon) or even continuous filament carbon fibre, and can therefore produce stiff but lightweight parts suitable for many applications in a production or prototyping setting, including fixtures. This technology produces more accurate and reliable prints than low-end desktop printers and, therefore, can provide a good transition from using 3D printing purely as a concept prototyping tool to one where the parts produced can be taken more seriously.

ENGINEERING TESTING FOR 3D PRINTING FIXTURES INCLUDES:

Accuracy: This is performed as a comparison to the 3D CAD model. This activity needs to be completed on different machines to ensure consistency in the model produced relative to the data input. There are many standard test models that provide data points for determining the accuracy of a print, with some allowing critical measurements to be taken with callipers and compared to the CAD model, and others providing more visual inspection such as the well-known "3DBenchy" [4].

Rigidity: This refers to the stability of the part. As this is critical for fixtures, the rigidity of prints needs to be rigorously tested during design development. Part locations will need to be held securely during operations, and the reliability of the print is therefore essential. Rigidity will be determined by several factors, including the material used, the additive manufacturing technology selected, the design detailing, the equipment settings, and the build orientation, as well as the settings for the print. For desktop 3D printers without an enclosure, for example, changes in conditions in a room will affect the integrity of the print, including temperature and moisture. 3D printers that have an enclosure can reduce the environmental impact of conditions on the part, but factors in a manufacturing environment such as humidity and vibration must still be addressed, particularly for accuracy and repeatability.

Durability: This will depend on how bespoke the fixture needs to be, as a one-off part may need to survive only a short-lived force, perhaps even a single use. However, a fixture may also be built from multiple parts to allow for adjustment where components are similar but not exactly a repeat part. It is possible to work with different additive manufacturing technologies to achieve different performance characteristics within a single fixture where appropriate. It is also possible to create multi-material parts for different performance characteristics, even with relatively low-cost filament 3D printers. For example, low-cost PLA could be used for the bulk of a fixture, while a more temperature-resistant material like ULTEM 9085 or PEEK (PolyEtherEtherKetone) is used in areas where heat is likely to build. However, this approach requires more advanced understanding of the technology and is not recommended for a company at the start of the adoption journey.

Delamination: For fixtures, this is a critical factor in a testing regime. It is also an opportunity to support the development of skills and understanding in the workforce. Trial and error, and gradual adaption of the design to reduce the delamination of a fixture, provides a good experiential learning context for the workforce in how to work with the technology. This can be explored using very low-cost desktop printers, then transferring the results to a higher-end printer, such as a carbon-fibre or industry-entry level dual-filament printer.

Design for support material: Even though the intent here is to work with additive manufacturing in an existing production context, low-level design for additive manufacturing (DfAM) awareness through designing for support material can be usefully introduced at this stage, as it is likely that the part itself will be straightforward enough for the effect of small changes to be easily seen and understood. Allowing individuals to learn by doing in this context is a useful approach, as it builds confidence in using the technology and the judgement of the individual, rather than being shown a finished part and told to memorise design rules. Whilst in many situations, DfAM is not necessarily intuitive, in this context it can be. Starting with simple fixtures and building the complexity is a good, long-term approach to professional development with the workforce.

Tensile testing: Testing for the tensile strength of the part, its yield strength, and its impact resistance (energy required to break a part) through this destructive test process, which measures the force required to break a part and the extent to which it deforms before breaking, is essentially the same for additive manufacturing as it is for other processes. However, the direction of printing must always be considered in the design of the test, and results will vary significantly based on print orientation [5]. Therefore, it is common for tensile testing of 3D printed samples to typically include three or more orientations – vertical, horizontal, and flat – although this will still lack information about different angles.

Shear testing: Similarly, a shear test, which measures the amount of shear stress necessary to deform or break a test piece, is as valid on additive manufacturing as it is on other processes, but only where the layer direction and design detailing are taken into consideration [6]. For this reason, multiple test pieces should be examined prior to the production of the final design, as detailing and layer direction, height, as well as build orientation will affect the outcome.

Wear testing: This is an interesting test because of the material used, the effect of temperature on the material, and, again, the laminations within the part. 3D prints are not isotropic (i.e., anisotropic); that is, its physical properties are not uniform throughout the part in any direction. Therefore, testing needs to be developed that reflects this [7]. This factor also impacts context-specific testing, such as heat resistance and vibration testing.

Grip test: One of the advantages of additive manufacturing that is frequently overlooked is the designer's ability to add surface texture to a part. This is a relatively easy process in additive manufacturing, compared to adding texture in a mould, for example. Different grip tests can be performed to evaluate geometries for desired applications.

One of the reasons the use of additive manufacturing for fixtures provides the opportunity to create transition parts is that the design does not need to mimic parts made using conventional production techniques. In fact, it should actively avoid mimicking conventional detailing, as such detailing is unlikely to maximise the performance of the part. Design detailing needs to be suitable for the technology in use, as outlined in the chapter on design detailing, as there are different detailing considerations for varying additive manufacturing processes. This, again, provides a good transition for workers in shifting from conventional manufacturing to additive manufacturing, as over-engineering the part is unlikely to be an issue in this context. It also allows the user to explore the impact of design detailing on the performance of the part. It is a low-risk, high-potential-reward approach to technology integration that sits between design and engineering and can utilise any 3D print technology.

JIGS

Repeating an action at an awkward angle or fitting a part onto an assembly accurately can be difficult to achieve. In addition to the consistency required, the effect on the worker repeating that action can be significant and may cause repetitive strain injury, particularly if the action involves holding the part to be fitted at an uncomfortable angle. Holding small pieces in very specific positions during fitting can be difficult. Jigs, used to support the fitting of large parts, can reduce the chance of injury or improve the accuracy of the fit. Well-known examples include jigs for the Hyundai Lifeboats, made using 3D printing through a collaboration with Materialise; and jigs for the Volvo production line, also produced by Materialise, where the weight of the conventional jig used was significantly reduced through 3D printing.

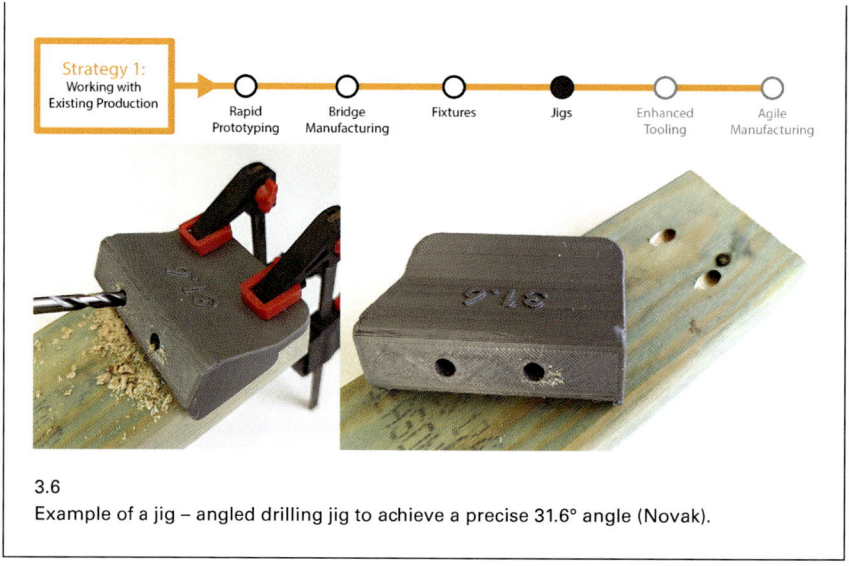

Strategy 1:
Working with
Existing Production

Rapid
Prototyping

Bridge
Manufacturing

Fixtures

Jigs

Enhanced
Tooling

Agile
Manufacturing

3.6
Example of a jig – angled drilling jig to achieve a precise 31.6° angle (Novak).

In production, jigs are used to hold the position of a tool (for example, whilst holes are drilled), to align multiple parts during a manufacturing process, or position details like stickers or badges. This provides the next-level opportunity for a product designer to introduce 3D printing in the factory through a positive, low-risk application, similar to the introduction of fixtures. Jigs are a low-risk introduction because they are unlikely to cause any consternation in the workforce; they can enhance the working conditions of individuals by improving safety and reducing ergonomic risks such as repetitive strain injuries. Jigs can be informed by the experienced and expert workers and can even be specific to an individual worker and their process. It is, therefore, an engaging product to work with and can be used at two levels during the integration of the technology: as a production tool, infiltrating existing practice; or in developing an understanding of the use of customisation. Beyond that, it can be used in personalisation, which is enabled by additive manufacturing, at a relatively low cost.

In comparison, mass production, or even batch production, implies a generic result for a predetermined number of a parts or products. Because workers on a production line are individual, their ergonomic needs are inevitably different, even for the same task. Yet, conventional manufacturing means that jigs will inevitably be, to some extent, generic. In addition, where adjustable mechanisms are built in to allow for some customisation, these add weight and complexity to the product. To create anything truly bespoke using conventional manufacturing would be expensive, if not impossible to source. With 3D printing it is possible to create a one-off, personalised product at low cost. For Strategy 1, more importantly, it gives ownership of the process to the individual, engages them in its design and production, and demonstrates practical benefit for the future. This is arguably one of the best approaches for upskilling the workforce on additive manufacturing and building confidence in its use. This provides a recommended precursor to introducing the technology into the production process itself. In focussing on this approach, product designers can work with individuals or groups to tackle ergonomic challenges on the factory floor on the premise that 3D printing allows for a bespoke solution not previously viable.

The same benefits exist for makers and artisans who often create jigs from leftover metal or timber to improve their workflow or accuracy for low-volume products. 3D printing can be used to begin digitising these supporting products as a maker learns CAD skills. The designs do not need to be particularly sophisticated but are the sort of thing that can be printed overnight in preparation for the next day of work. This may improve accuracy compared to fabricating a jig manually. For example, in CAD it is easy to create a drilling guide that is at exactly 31.6° (as shown in Figure 3.6) – a level of accuracy virtually impossible with a handmade jig. As described in the Fixtures section, designing jigs parametrically will mean that a maker, or a company, will build a library of templates over time that can be slightly adjusted and used many times over, shortening the process to create jigs.

Whilst the intent for this approach is predominantly about upskilling the workforce, creating effective jigs is still key so that the approach does not backfire, undermining confidence in the technology. For this reason, professional product designers should be involved and the products tested prior to use. Further, they should, in collaboration with workers, identify a suitable task to accomplish. An example would be where parts must be fitted to an assembly in an ergonomically awkward situation, such as kneeling or stretching, but where the part still must be fitted accurately, a common occurrence in the automotive industry. A jig can be designed to improve ergonomics and help quickly align parts, so a worker's risk of injury is dramatically reduced, while also improving efficiency of the production line. Another example would be fabricating a large one-off public sculpture where a single maker is trying to attach bolts through large panels but cannot reach the other side to hold the nuts in place. Without calling in someone to assist, a 3D printed jig could be produced that sits on the edge of the panels with nuts slotted into cavities corresponding with the holes. The maker can simply thread the bolts through all together and release the jig prior to tightening. Such outcomes are measurable in both the factory and workshop setting and can be evaluated over time, building a strong case for the adoption of 3D printing in the company.

An alternative way to identify a task is to work with any concerns the workers have in relation to injury prevention or compensating for an injury. For example, a sports injury may be relatively temporary, but it could make physical work more difficult during recovery. Wherever a personalised solution is needed, such as with ergonomic jigs, generating accurate data can be critical. 3D scanning has evolved alongside 3D printing over the last two decades and, for product designers, is becoming an essential tool in partnership with the technology. In terms of workforce engagement and upskilling, scanning has the bonus of a novelty factor. At the time of writing, it is still unusual to be professionally scanned, and it can therefore be a useful way of engaging anyone reluctant to be involved. Much as with 3D printing, there are different levels of scanning tools and software that produce very different results [8]. For this stage of the adoption "timeline," it would most likely be useful as an engagement tool rather than a practical one as it can be difficult to use accurately. Like additive manufacturing, it takes a level of skill and understanding of the software. Scanners, like 3D printers, are available at a wide range of prices and use different technologies. From scanning apps, which can be problematic, through handheld scanners that cost hundreds to those that cost thousands, the type of scanner and sophistication of its software and use will affect the outcome. This is very much a specialist skill, much like DfAM, but works very well in tandem with the fabrication technology.

Whilst scanning may not be needed during the early stages of additive manufacturing adoption in a workplace, it is increasingly seen as an essential tool for product designers in the use of digital technologies. This becomes apparent in the next chapter, where product redesign is discussed. Learning to use a scanner, and then integrating the technology into workflow, provides product designers with an additional skill that will be useful as part of the suite of digital technologies converging, at this time, to provide new opportunities for the profession.

KEY POINTS

Lowering the barriers to 3D printing adoption:

- This is a suitable justification for the introduction of low-cost desktop printers into a company. This approach involves a small financial investment by a company, and a very low level of risk.
- Workers can gradually become familiar with 3D printed parts and, once some jigs have been implemented in the facility, may proactively identify new opportunities for 3D printed jigs or other products to streamline processes.
- For the product designer, jigs can be designed alongside the parts in CAD. Therefore, a jig does not need a large additional time/financial investment on top of designing the part to begin with.
- If the jig breaks, or the design changes, it is possible to 3D print a replacement using the same CAD file and minor changes.
- While digital warehousing of jigs is possible, in some cases, it may be more affordable/practical to recycle a jig when it is not being used than pay for storage space. A new one can be 3D printed if needed again and will be the same as the original.

New opportunities to streamline manufacturing:

- *Increased productivity with reduced scrap/waste*: Using jigs is known to reduce the opportunities for errors while making/assembling a product. Therefore, there is less risk of error and less product being thrown away.
- *Increased accuracy and repeatability*: Jigs can improve tolerances and other aspects of a product, meaning that parts may be more reliable.
- *Reduced cost compared to machining a custom jig*: While jigs are useful, many times they are not used because of the cost to manufacture one. Because of the low cost of FFF 3D printing, jigs can be affordably produced, and many more could be implemented within a facility to streamline production.
- *Increased assembly complexity possible*: Jigs allow numerous parts to be located together simultaneously and assembled in a more efficient manner than not using a jig. This may also mean fewer workers are required to create a part or assembly.
- *Suitable for a one-off use or repeated use*: A jig may enable customised production due to the affordability of producing one, opening new opportunities to allow a product to be modified to suit individual customer needs.
- *Speed*: For manufacturers who do not have machine workshops to fabricate their own jigs, a 3D printed jig can be manufactured in-house faster than one that needs to be outsourced. This increases speed to market and agility.
- *Better accuracy and reliability*: Through newer colour 3D printing, for example HP MJF, instructions and other information may be directly 3D printed as part of the jig, further improving accuracy and reliability.

Improving worker health and productivity:

- *Increased safety for workers compared to not using a jig*: By supporting parts during production or assembly, there is less risk of things going wrong when drilling, cutting, or welding.
- *Ergonomics*: A jig can assist how parts are manoeuvred or assembled, ensuring repetitive tasks are performed safely.
- *Workers with lower skill levels may be able to better perform a task*: A jig with guides for drilling or cutting may mean that workers can more accurately create high-value parts without the same level of experience formerly required to make the same part without the aid of a jig.

Considerations for designers:

- Design the jig at the same time as the part/assembly in CAD.
- 3D scanning can be utilised to create jigs for parts that do not have appropriate CAD files.
- Hybrid fabrication of jigs may be necessary; for example, areas prone to high wear may need to be metal whereas the rest of the jig may be polymer or composite.
- Using inserts which are heat welded into 3D printed parts can add a lot of extra functionality. For example, threaded inserts mean that screws can be used to hold or attach parts during assembly.

ENHANCED TOOLING

The example of working with existing production to enhance tooling, shown in Figure 3.7, demonstrates how additional cooling channels can be designed into a moulding tool to speed up the production cycle because the mould and parts cool more quickly. This method allows for complex geometry that follows the shape of the part to be designed into the mould, which would be extremely difficult to produce any other way.

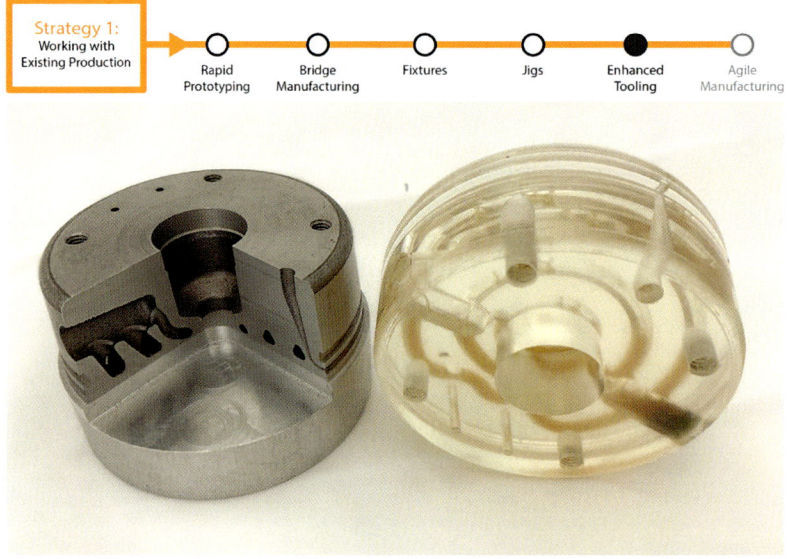

3.7
Example of enhanced tooling – conformal cooling channels shown through both cross-section (left) and transparent print (right) (Diegel).

For a company that is heavily invested in conventional manufacturing processes and that has an established workforce, another effective way of introducing the technology into the company and building knowledge on additive manufacturing processes is to look at enhanced tooling as the entry point. Additive manufacturing allows for enhancements in tooling because it is not constrained by many of the limitations in making tools for conventional manufacturing. Critically, fabrication can be made to follow the design requirements, rather than the other way around. That is, the optimal geometries can be used to form the basis of the tool design rather than designing to the practicalities of how milling or casting are used to create production tools, such as moulds. Essentially, this means that with additive manufacturing, the tool can be redesigned from the "inside out." Straight lines can be replaced with organic curves, and internal channels can be incorporated into the design. This is particularly useful in the example of designing tools to include cooling channels.

Conformal cooling, for example, where a cavity is added to a mould core in a shape that follows as much as possible geometries in the mould (either of the part being made or the geometry of the mould itself), is one such approach. Basically, it conforms to the critical shapes in the tool. Conformal cooling is added to improve cycle times, part quality, and tooling life [9]. The concept of conformal cooling is not new to manufacturing, but typically it has relied on drilling straight holes through a mould that intersects with entry and exit holes plugged to create a closed loop for cooling fluid. This drilling process means the channels may not be placed in the optimal location for a part due to the limitations of drilling straight holes. It also means the pathway is not as efficient for the fluid. With additive manufacturing, the cooling channels can be more accurately conformal, shaped around the curves of a part in any direction, and in three dimensions.

In additive manufacturing, typically metal powder bed processes are used for this application. Because of its layer-by-layer process, internal cavities in complex arrangements can be formed. This means cooling pathways can be placed equidistant from a part's surface to reduce the build-up of stresses as a part starts to cool. Alternatively, their density can be increased where necessary to help cool a part in a particular area. This flexibility also means the pathway for cooling fluid can be more efficient. Different arrangements of the cooling channels can be used for different effects; for example, it is possible to cool the part, in order, from one end to the other, or to cool the whole part at the same time. Simulation software can be used to help prepare moulds and analyse the predicted results of cooling methods.

Additive manufacturing allows advanced tooling to be designed that goes beyond what was possible to produce with conventional tooling. Tools that would be built from multiple parts can be made as a single part, and complexities that would be difficult to produce, and reproduce accurately, can be integrated into mass-production practices. An example of this is the tooling for truck tyres produced by global manufacturer Michelin. Michelin introduced 3D printed metal stipes in their moulds in 2013. This tooling enabled the tyre tread to incorporate complex patterns that changed the grip of the tyre as it wore down during use. This example illustrates excellence in DfAM as it exploits the characteristics possible with the technology to add value for the company. At the same time, it is working with existing production practices and utilising their understanding of additive manufacturing technology and tyre design to enhance their in-house capabilities.

The complexity of these types of tooling means that they would be difficult to produce using conventional techniques. This is a sophisticated approach, and an indication that a company is becoming more experienced with the technology and has engineers who have a good working knowledge of the technology. To be able to envisage this type of proposal, there must have been a good level of knowledge of working with metal additive manufacturing. The scale of the part requires a considerable investment to be made. Tooling can also be produced from polymers, particularly for low-volume production. This is covered in the following section.

AGILE MANUFACTURING

Because of the relatively low cost of producing one-off items with 3D printing, it is possible to create moulds for a process such as vacuum forming, as in the example in Figure 3.8. This approach allows for a highly adaptable product range, enabling a company to respond to the market rapidly and keep their product range fluid.

3.8
Example of agile manufacturing – a one-off mould for vacuum forming utilising a 3D scan of someone's face, to be used for some chocolate moulds (Novak).

In this approach, the capabilities of additive manufacturing to create one-off, or short-run, tooling parts are exploited to build the ability of a company to respond more quickly to market forces. This ability has been highlighted as a particular need during the recent COVID-19 pandemic as supply chains were disrupted and demand for specific products in short supply rose dramatically. During this time, companies that were able to pivot supply towards these new temporary markets were able to take advantage of the profits created by the immediate demand. In addition, the disruptions to the supply chain affected imports of critical parts and materials, as well as the ability of a company to export its products. Companies that were able to quickly adjust their products to meet these challenges were best positioned to survive.

Even without these short-term disruptions, the ability to adapt to changing market demands is increasingly important for a company to future-proof itself against global and local shifts in demand for items, as well as burgeoning competition. Tooling made using additive manufacturing may not be cheaper than conventional tooling if it mimics the existing part. However, it is possible to create the tool with a shorter turnaround, which can provide a company with a much more valuable market advantage. This can be achieved in different ways. For example, a mould could be developed that has a conventional exterior shell but has a short-run insert that is additively manufactured. This would allow the company to produce more targeted products. Essentially, the intent is to create temporary tooling that can either work in conjunction with existing mass-production tooling, or even replace those tools for a highly responsive production line. This is less likely to happen in a large company but could be quite useful for an entrepreneur seeking to gain an edge in the market.

Short-run moulds

Polymer moulds have been used for short-run production for many years in traditional manufacturing. In lieu of CNC machining, additive manufacturing can be a cost-effective method of producing this type of tooling for a short run of parts. This application uses polymer 3D print processes, although different post-processing may add different materials to improve wear and surface finish. Selecting the most appropriate additive manufacturing technology depends on the application and material being formed in the mould. The capability of additive manufacturing for integrating textures into a design with relative ease, such as on the inside of a mould, is arguably under-utilised. Textures can be added to any surfaces, internal or external, that traditional mould-making methods cannot achieve. Additionally, assemblies may also be produced as a single part, subject to the limitations within additive manufacturing itself.

Hybrid tooling

Hybrid tooling refers to the combining of traditional tools with a component that is 3D printed. This may be to accommodate conformal cooling, only necessary in a complex area of a mould, or as a custom insert to provide a flexible element in the mould, for example, for custom branding or a short run of product. Custom inserts can be 3D printed in metal or use SLS backfilled with a metal like copper. Maintaining the shell mould and replacing the inserts with 3D printed parts allows the company greater flexibility in the parts they can produce as short runs, without the need to invest in expensive tooling each time. Large, solid forms are rarely worthwhile when produced using additive manufacturing, particularly when made from metal which is expensive, and there are significant limitations on the size of prints possible. Hybrid solutions arguably maximise the capabilities of conventional and additive manufacturing technologies. This approach, as well as the use of hybrid additive manufacturing technologies, is an area of growing interest.

Casting

There have been interesting developments recently in casting using 3D printed moulds in terms of both scale and range of application. Large-scale, direct 3D printing of sand moulds for casting using a binder jetting system reduces the cost of casting and is used increasingly in construction. Very large 3D printers for casting are part of developments in large-volume 3D printing, led by research organisations such as Oak Ridge National Laboratory in the United States. In contrast, wax 3D printing is used for lost wax casting for fine detail jewellery. SLA and DLP are also used in casting for jewellery. These processes allow for a high degree of complexity in design with a good surface finish. The use of additive manufacturing to work with established production practices has been embraced by high-end jewellery manufacturers as well as individual makers, since good-quality SLA and DLP printers are now similar in price to desktop FFF machines.

Vacuum forming moulds

This is the process commonly used now to manufacture clear dental aligners. SLA/DLP, SLS, and MJF are all good candidates to manufacture moulds for this type of application. 3D printing is also used to create vacuum forming moulds for packaging, as well as bespoke signage. While filament printing can be used, post-processing may be required to create smooth surfaces; otherwise, layers can appear in the formed piece. Midscale 3D printing moulds for forming processes such as the vacuum bag for veneer lamination can also be made, though it is worthwhile only if there is a level of complexity in the mould that is difficult to achieve using conventional workshop practice. This approach can also be used for composite forming moulds such as for carbon fibre. This has become more viable recently as midsize filament printers, approximately one square metre in size, have become more accessible, as described in the furniture section in Chapter 6. In addition, more individuals and companies are building their own 3D printers that can be used for midscale prints, such as in the example of Studio Kite in the film section of Chapter 6.

The degree to which the use of agile manufacturing is adopted depends on the level of ability in working with additive manufactured moulds within the company as much as on external markets. Disruptions to supply chains, which lead to short-term opportunities, may create the need for a more responsive production cycle. Even where a company practices this approach as a precautionary, future proofing measure, it will serve to enhance in-house knowledge and skills. In the agile manufacturing approach described here, the term "rapid tooling" is sometimes used to refer to the temporary (short run or one-off) or supplementary tools used [10]. Whilst numerous forms of rapid tooling are already in use in production in conventional manufacturing, additive manufacturing allows for a greater range of adapted tooling options, in this context, and can also reduce the costs depending on the circumstances (e.g., 3D printing moulds for very short runs). A product designer may propose this approach for a company to try as a knowledge-building exercise without disrupting everyday practice. This type of future proofing will help change thinking about the technology and its application in a business sense for the company.

Whilst this adapted tooling approach could be used as a precursor to end-use batch production, it would require a rethinking of the product for that purpose, or maybe the development of new, more suitable, products beyond the current offerings. Therefore, it has been included in Strategy 2 – Product Redesign and New Product Design – instead of here. This approach allows a company to be more experimental with products, a concept further discussed in Strategy 2. Nevertheless, in terms of working towards building technology adoption within a company, the added value is in building acceptance of the technology in-house by supplementing rather than replacing any of the existing production lines or devaluing current skills. For this stage of technology adoption, constraining an adapted tooling approach to work with existing production practices will avoid the complexities of business model change required for a complete shift to production using additive manufacturing. Therefore, it supports positive adoption of the technology by the workforce with less risk. The following are summaries of common agile manufacturing methods where additive manufacturing can be used to integrate within existing practice and build new capability.

CONSIDERATIONS FOR DESIGNERS:

- 3D printed tooling may need to be designed with extra material so that the rough surface from 3D printing can be smoothed by machines as a secondary process. As a guideline, an extra 0.5 mm in most areas where machining is required should be enough.
- Consider whether 3D printed tooling is the best method (see the section "Bridge Manufacturing" earlier in this chapter), or whether other alternatives exist which may also achieve the desired outcomes, particularly for low-volume manufacturing.

- Product designers frequently have a choice between direct 3D printing and in-direct manufacturing; for example, using lost wax casting for bronze casting or 3D printing directly in metal. The outcomes produced can be quite different and affect the detailing – for example, wall thickness.
- Large moulds – for example, for veneer forming for furniture – can be 3D printed via large-scale FFF printers. However, the time taken to print and the volume of material for printing need to be considered. Where possible, adding in hollow areas to large blocks and designing to incorporate support structures, rather than allowing the slicing software to generate them, can create efficiencies.

BUILDING A STRONG FOUNDATION IN ADDITIVE MANUFACTURING THROUGH STRATEGY 1

Working through these approaches to introducing additive manufacturing into an established production facility, either all or only the ones most suitable to the company, should provide a good foundation for teaching basic skills in additive manufacturing, and familiarity with the tools involved. It is hoped that they provide a pathway to upskilling the workforce without causing too much disruption in the first instance.

In terms of workforce development, it is also recommended to invest time in running workshops or presentations at different levels within the company, not only with workers in the production line. The intent is to raise awareness of the technology and the opportunities and disruptions it can cause to all stakeholders, rather than focussing solely on the factory floor. This aspect of workforce and organisational development is overlooked, but for the long-term use of additive manufacturing as a driver for innovation, it matters to invest in upskilling across the board. If decision-makers have a superficial understanding of the technology, it can be used inappropriately, or dismissed as merely a workshop tool.

DESIGNER'S COMMENT

The complexities of manufacturing make it difficult to test ideas. 3D printing removes this barrier and allows ideas to easily, and quickly, be tested. This makes 3D printing a great catalyst for innovation.

Further, for Strategy 1, the focus should be on business efficiencies wherever possible, from production to human resources (workforce morale, reduction of workplace injuries), and how they can be supported by decision-making within the organisation. Whilst the focus here is on additive manufacturing, this could form part of broader professional development on disruptive digital technology and serve as an introduction to integrating "the digital" in an established manufacturing plant.

For some companies, integrating additive manufacturing into existing practice may be the most efficient way to use the technology long term. Moving on to Strategy 2, focussed on product redesign and innovation, provides the potential for high-profile change but also has significant implications for a business, and therefore, it needs to be thoroughly explored and understood for the company before introduction. Working through the layers of Strategy 1 with a company should help to build trust in the processes and the designers' ability to lead the company through their integration, prior to the more disruptive strategies identified within Strategy 2 on product redesign and new product design, and eventually Strategy 3 on changing business models. Difficult though it is to resist jumping into product redesign or design innovation when working with a company, building that level of trust and knowledge in a company first can help the successful adoption of additive manufacturing in a company long term. Additive manufacturing encompasses a complex range of technologies that can be difficult to navigate successfully, but even more so, it can involve complexities in workforce development and systems organisation that can create change management issues. Recognising these challenges and working to mitigate the problems they could cause should establish their long-term value and lead to buy-in, skills development, and positive innovation, over problems.

MEANWHILE, WHAT MIGHT HAVE HAPPENED IN THE FAMILY FIRM?

In the hypothetical scenario of Madoc Manufacturing, they decided to follow LND Consulting's recommendations and formed a small working group to improve ergonomics on the production line and in the warehouse, and look at new technologies to do this. They brought in an outside expert on work safe handling, and Erin worked with the ergonomics expert and Zac from LND to identify where 3D printing could be useful as part of the recommendations being made. Erin and Zac focussed on developing jigs and fixtures to reduce some of the health and safety issues caused by the nature of the work, utilising a local service bureau operating out of a university to produce the first couple of designs.

The workforce saw the move as positive, and additive manufacturing as a useful tool, rather than a concern. Erin recommended to the family that these workers form a 3D printing team to identify ongoing opportunities for design to support operation in the workplace. The company invested in two desktop printers and a carbon-fibre printer. The original ergonomics working group voluntarily became a 3D printing working group, several of them opting into an online course to learn CAD.

Over the following months, they gained confidence operating the 3D print-ers, giving demonstrations to others in the factory. They started with a relatively straightforward use of the technology with their first 3D printed jig shown in Figure 3.9. Lewis pointed out that the part could have been made using laser cutting. However, Harry and Sara saw it as the first small step in using the tech-nology. Over the next six months, the whole team gained confidence in the process – enough to decide to attend the next Manufacturing Week event with the express idea of looking at industrial machines and the opportunities they might bring. The following year, the toolmakers, led by Erin and a surprisingly enthusiastic Lewis, designed and 3D printed test prints for moulding tools incor-porating conformal cooling channels, which they then had 3D printed in steel through a service provider.

On to the next stage of the journey …

3.9
Example of a
positioning jig
(Novak).

REFERENCES

1. Gibson I, Rosen D, Stucker B. *Additive Manufacturing Technologies: 3D Printing, Rapid Prototyping, and Direct Digital Manufacturing*. 2nd ed. New York: Springer; 2015.
2. Novak JI, Loy J. Moving 3D Printing beyond the Desktop within Higher Education: Towards a Service Bureau Approach. In: Ali N, Khine MS, editors. *Integrating 3D Printing into Teaching and Learning: Practitioners' Perspectives. Contemporary Approaches to Research in Learning Innovations*. Leiden, Netherlands: Brill; 2020. pp. 206–227.
3. Camba JD, Contero M, Company P. Parametric CAD Modeling: An Analysis of Strategies for Design Reusability. *Computer-Aided Design* 2016;74:18–31.
4. Kim GD, Oh YT. A Benchmark Study on Rapid Prototyping Processes and Machines: Quantitative Comparisons of Mechanical Properties, Accuracy, Roughness, Speed, and Material Cost. *Proceedings of the Institution of Mechanical Engineers* 2008;222(B2):201–215.
5. Torrado AR, Roberson DA. Failure Analysis and Anisotropy Evaluation of 3D-Printed Tensile Test Specimens of Different Geometries and Print Raster Patterns. *Journal of Failure Analysis and Prevention* 2016;16(1):154–164.

6. Rohde S, Cantrell J, Jerez A, Kroese C, Damiani D, Gurnani R, et al. Experimental Characterization of the Shear Properties of 3D–Printed ABS and Polycarbonate Parts. *Experimental Mechanics* 2018;58(6):871–884.
7. Pant M, Singari RM, Arora PK, Moona G, Kumar H. Wear Assessment of 3–D Printed Parts of PLA (Polylactic Acid) Using Taguchi Design and Artificial Neural Network (ANN) Technique. *Materials Research Express* 2020;7(11):115307.
8. Daneshmand M, Helmi A, Avots E, Noroozi F, Alisinanoglu F, Arslan HS, et al. 3D Scanning: A Comprehensive Survey. arXiv. 2018.
9. Jahan SA, El-Mounayri H. Optimal Conformal Cooling Channels in 3D Printed Dies for Plastic Injection Molding. *Procedia Manufacturing* 2016;5:888–900.
10. Equbal A, Sood AK, Shamim M. Rapid Tooling: A Major Shift in Tooling Practice. *Journal of Manufacturing and Industrial Engineering* 2015;14(3–4):1–9.

4 Strategy 2

Product redesign and new product design

The second stage of the strategic adoption of additive manufacturing into a company, driven by product design practices, is described here as "Strategy 2: Product Redesign and New Product Design," where the focus is on the potential of additive manufacturing to add value. When a product designer has worked through the different approaches outlined in "Strategy 1: Working with Existing Production Practices," and the company has built up sufficient skills and experience, as well as enthusiasm, for taking the integration of additive manufacturing to the next level, it is then possible to move onto Strategy 2. Strategy 2 moves a company on from enhancing conventional manufacturing practices, and building worker experience with additive manufacturing, to the next stage of technology adoption. This strategy focuses on designing and producing end-use products to sell to the customer (Figure 4.1).

4.1
Overview of Strategy 2 approaches to 3D printing.

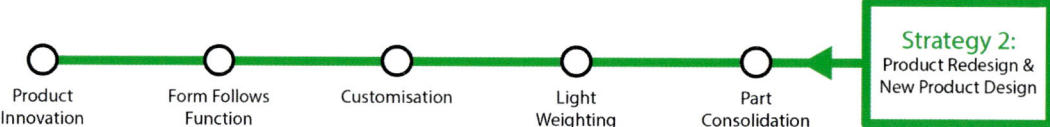

THE SME: AN AMBITIOUS MIDSIZED COMPANY LOOKING TO RETAIN THEIR EDGE IN THE MARKET

Fox Coral Solutions is a fictional midsize company that has shown rapid growth over the last decade in the design and production of high-end automotive after-market products. It was founded by Ali Khatri and Siva Annand, who met at university whilst studying business. They took considerable risks to push forward an ambitious agenda, determined to be leaders in product design and fabrication for their client base. They built a company of around 200 workers and, wherever possible, they brought in technology systems to help automate production and drive the brand forward. The company has an in-house product designer, Liam Bakker, who has considerable experience in design for manufacturing, and who has successfully built a product range that caters to the current market. However, Liam is concerned that motorcycle after-market products are a competitive market, particularly due to cheap imports from

DOI: 10.4324/9781003122203-5

4.2

overseas, and is keen to add to the company's capabilities. He discusses his concerns with Ali and Siva. Siva is very enthusiastic about the idea of expanding their current use of 3D printing from rapid prototyping to end-use products. Liam and Ali are both interested in the idea but unsure what the move would entail and where to start. They decide to bring in a design consultancy with expertise in design for additive manufacturing to advise them. They call LND Consulting (Figure 4.2).

Meeks visits the company, armed with samples and full of ideas. She is already familiar with the work of Fox Coral Solutions and sees it as an opportunity to demonstrate the impact that additive manufacturing can have on a business in the right environment, with the right people, and with the right strategy. Ali shows her around the factory, where Meeks talks to the workers. They are all familiar with desktop 3D printing, having two desktop machines that are used for fabricating prototypes and jigs, and are all comfortable with digital production techniques. Production manager Noah Jans, a long-time friend of Liam's, is also enthusiastic but wants to understand how to shift their practice from desktop machines to those for end-use commercial production. He shows Meeks examples of parts he had printed using a service bureau and asks her how difficult it would be to bring a similar industrial machine into the factory. Noah joins Liam, Ali, and Siva in the meeting room for Meeks to talk through the examples she has with her, illustrating the different stages in integrating the technology into the practice. Meeks recommends that the company work through Strategy 2 as a way of building the in-house expertise they need to integrate the technology. She explains her reasoning and how it would work.

WHY WORK WITH STRATEGY 2?

This chapter is for product designers working directly on products for market with companies at a similar stage to that of Ali and Siva's company. That is, the company has experience using 3D printing for prototyping and for in-house products, such as jigs and fixtures, and they have in-house expertise, in this case Liam, who understands how to design a part appropriately to ensure it does not fail during the print, but not where to go next. This is the next stage of technology adoption beyond its use in production line optimisation, and building on initial workforce development in accepting and using the technology.

This step is recommended for companies that have built that foundational knowledge of 3D printing and are equipped to take the next step towards integrating the technology into practice – particularly where they are ready to invest in the technology and build their in-house, technical, and design for additive manufacturing (DfAM) expertise. It is not recommended as a starting point for companies totally new to the technology, as considerable challenges and commitment are involved in taking the steps outlined here. However, in addition to providing the next steps for companies who have completed Strategy 1 (or the equivalent), it provides direction for the product designer when consulting with a company which has decided to jump straight into new product design or redesign. This includes start-ups and designers who may be able to leverage additive manufacturing from the outset.

This chapter includes advice on how to work with companies to prepare them as much as possible for the design, production, and distribution of commercial 3D printed products. For Fox Coral Solutions, understanding what products they could make using additive manufacturing, how to start the process, and what it will involve is an exciting place to be. Noah and Ali meet regularly to discuss how much it will cost and what could go wrong. Liam, meanwhile, daydreams about future design competitions and how their company will make its mark in the industry.

DESIGNER'S COMMENT

A golden rule of additive manufacturing: It depends! There is never a definitive answer with this technology or a one-size-fits-all approach. The answer to almost any question depends on many factors.

Strategy 2 includes the most visible areas of DfAM for the product designer. The approaches discussed here include building on existing products to add value, and product redesign, as well as product design innovation, to truly exploit the opportunities that additive manufacturing provides. These are the aspects of DfAM that tend to feature in magazines and win competitions. However, for sustainable commercial products to be created for a company, which can be effectively integrated into production practice, there needs to be considerable work behind the scenes.

Although there are many design rules for working with additive manufacturing, a different mindset and a considerable body of knowledge are needed when working with the range of technologies available. This imperative is like working with other emerging technologies. Designing a product for additive manufacturing that sits within the portfolio of a company may be relatively straightforward for a consultant product designer with a specialisation in the technology. It ensures the product can be produced in the long term, and without too much disruption or risk. This chapter provides direction and examples for product designers working on new products and enhancing existing products that utilise the opportunities that additive manufacturing provides. It builds on the step-by-step approach for integrating additive manufacturing into production practice introduced in the last chapter. Again, the approach is to work from the least disruptive and arguably most straightforward, to the most innovative ways of working.

TECHNOLOGY ADOPTION JOURNEY MAPPING: PRACTICAL CONSIDERATIONS

Ideally, a company will have worked through Strategy 1, with or without a product designer. If not, there may be a specific reason that a company needs end-use parts to be designed for additive manufacturing. Either way, they feel ready to take the next step. For the most part, this is the stage of technology adoption in which product design consultants specialising in additive manufacturing operate across numerous companies.

Through this strategy, new products that maximise the relative geometric freedom of additive manufacturing are developed. It is the showcase section, where product innovations enabled by the technology are most easily visualised and understood by the public. Yet, this stage can be fraught with problems. A company that is not prepared for the costs and commitments involved, a workforce that is resistant to the changes the technology creates, and a supply chain that has not been thought through for the technology, all add risk to the venture. For a successful launch, these issues need to be considered seriously, with all the pros and cons for that company understood as part of the process. The design of a part or product for additive manufacturing in isolation is unlikely to be a wise investment.

As well as new opportunities, the hype around 3D printing has created challenges for product designers because of unrealistic expectations and a lack of awareness of the pitfalls of the technology. This is also a reason why a product designer over an engineer is needed. This owes to the product designers' role generally involving a more holistic view of production and the market, whereas engineers tend to be focussed on solving narrowly defined, technical problems. The business implications for the adoption of even a single part using 3D printing can be significant for a company, and understanding and working through those implications before making a commercial commitment can make the difference between the part being a short-lived curiosity and one with longevity in the product catalogue. Engineers tend to have more of a focus on the mechanical considerations in production and prefer working with specifications and measurable qualities. Ideally, the product designer and engineer work together to create a viable and commercial solution.

Many issues need to be taken into consideration for any company looking at investing in industrial additive manufacturing machines and producing end-use market-ready parts on a regular basis. These tend to be more important for end-use products rather than for the supporting products outlined through Strategy 1 and include, for example, rethinking material source, material handling, material storage, material disposal, and recycling.

Material sourcing: Conventional material suppliers are unlikely to currently stock materials suitable for additive manufacturing. New providers will need to be sourced, and the material validated. This can be a problem because much of the material supplied for 3D printers can be proprietary, which means there needs to be a level of trust in the data supplied by that company on the performance characteristics of the material. In addition, how material performs once it has been through an additive manufacturing process may not align to how it performs going through a more conventional process. This means the company is essentially starting over in terms of material knowledge.

Material storage: Many of the polymer filaments are hygroscopic. This means that for some of the machines, the filament arrives in sealed packs, and once the seal is broken, the filament should be used relatively quickly and stored in a low-humidity facility in the meantime. Powdered materials can be even more challenging, as the characteristics of the powder could affect the outcome of the process. In addition, some metal powders can be explosive if not stored in appropriate conditions, and often need to be kept separate to the main production facility.

Material handling: Material properties are very different for the varying types of additive manufacturing available. Understanding the drawbacks with different materials in terms of handling is critical prior to committing to a specific type of additive manufacturing technology. For example, stereolithography, known as SLA, is a resin-based process. During the process, the resin gives off unpleasant fumes that can impact a large area, even when standard extraction is used. For this reason, a dedicated facility with industrial extraction is recommended where SLA machines operate. Additionally, many of the most common resins used are not safe for skin contact and can cause irritation or worse without appropriate PPE.

Material disposal and recycling: The disposal of filaments is, on the surface, quite simple. They are the same plastics used in common products and packaging and are, in theory, recyclable. An increasing variety even have biodegradable qualities. However, this is ultimately a waste product that is not easily recycled, and large quantities of plastic can be discarded in the form of unused/spoilt filaments, as well as failed prints, support material, and prototypes or other parts that are no longer needed. Long term, this waste needs to be addressed and, while recycling solutions that shred waste and turn it into new filament can be purchased, these rarely result in quality materials that can be reliably used again. Powders have a better chance of multiple use, with a percentage of recycled powder combined with virgin powder for use in most powder bed fusion systems (polymers and metals). Some suppliers provide return and recycle opportunities for a company, but this is not yet available worldwide because markets remain relatively small in certain countries. There are also issues with the disposal of used resins in SLA that need to be addressed, as well as purged resins from material jetting.

The cost of industrial machines for production is, not surprisingly, significantly higher than the desktop machines. Dual-filament, industrial-grade machines start at approximately seven times the cost of a good-quality desktop, for example, and the filament costs are also considerably higher, typically constrained to a particular manufacturer through electronic chips that prevent a generic material from being used. In addition to the cost of the machines themselves is the cost of installation and ongoing maintenance. This can be a serious investment for a company depending on the technology chosen. For example, metal powder bed printers can weigh more than a standard floor can support, requiring the area to be re-enforced. Extraction for these machines is a critical consideration since breathing in the powder is a serious health and safety issue, and different gases can be used and produced during the printing process. Planning the installation of such a printer in a basement workshop may, therefore, not be the best idea. Even desktop FFF machines give off fumes at varying levels, depending on the filament used, and should ideally be extracted [1, 2].

Once the parts are out of the 3D printer, they will require post-processing – again, this is different for different technologies, but anything that involves powder will require a finishing station with appropriate extraction to prevent the powder from spreading throughout the facility and affecting workers and other machines. The floor space necessary to provide clearance around industrial 3D printers also needs to be catered for. These costs can more than double the price of installing an industrial 3D printer and, thus, need to be researched thoroughly before a company commits to a purchase.

Another issue for any company considering buying industrial additive manufacturing machines is finding the suitable technicians to operate and maintain the machines and, more importantly, know enough about the process to maximise the output. In addition, the hands-on, frequently messy nature of post-processing could dissuade technicians from working solely with additive manufacturing machines.

All these are factors, even before considering the cost of building skills in the workforce, developing new product, and understanding the impact of the technology on distribution and product lifecycle. Material handling, changing the very nature of production, upskilling and displacement of the workforce, the cost of industry-level machines to buy and maintain, and the changed consumer interactions are among some of the key considerations of products manufactured using additive manufacturing. They need to be mapped prior to any investment. On top of that, selecting the appropriate technology, designing effectively for that technology, including for post-processing, requires knowledge and experience. That said, the opportunities the technology provides are also significant and well worth the trouble, if handled carefully.

Skipping over the first strategy to this more challenging second strategy can be a mistake which may cause resistance from the workforce as the challenges involved emerge. This ultimately results in problems with support for the inclusion of the technology on the shopfloor. In addition, a company is unlikely to have the operational support system in place to deal with a product redesign. This includes from a technical point of view, as well as equipment, logistics, and customer demand. Thus, the strategy described herein is appropriate once a company is ready to consider product redesign, as the strategy provides progressive approaches to integrating the technology into practice.

As emphasised throughout this book, the three consecutive strategies outlined do advocate for constraints on the product designer in terms of what is proposed to a company and when. It is tempting for a consultant to provide a dramatic, eye-catching solution when asked to work in 3D printing by a company new to the technology. However, for the long term, working within the limitations of the technology adoption by that company should create more successful outcomes. Just as this approach advocates for completing the first stage of adoption with the company outlined in Strategy 1 before moving on to product proposals, so too it advocates for keeping the reigns on changing business practice too radically before Strategy 2 is also complete.

Strategy 2 focusses on product redesign and innovation that builds on a foundation of skills and knowledge of the technology in a company without proposing a whole new way of interacting with customers and a radical shake-up of the company. Tempting though it is to leapfrog the earlier strategies, given the expectation that the quicker a company can move to contemplating Strategy 3, the more able it will be to respond to 3D printing and subsequent disruptive digital technologies, the company needs at least a degree of the unpinning skills and knowledge of both Strategies 1 and 2 before progressing. Upon the completion of Strategy 1 in this book, if the workforce and executives in the company have an informed, positive view of 3D printing, then the professional product designer can lead them onto product redesign. Strategy 2 starts with the least intrusive application of the technology for end-use products, which is the technical optimisation of existing products; it then works through light-weighting, topology optimisation, and part consolidation for product redesign; and then on to more complex approaches to integrating the technology for new product design.

With all the challenges involved, transitioning an existing practice to this stage of technology integration, or setting up a new company that creates all or some of its product from additive manufacturing from the start, if approached carefully, can have a major impact on the products a company can produce as well as enhance the image of the company as forward thinking. The marketing opportunities that arise can offset some of the inherent costs of adopting the technology at an industrial level. The transition just needs to be done correctly. The stages suggested for Strategy 2, outlined here, build from engineering optimisation and personalisation strategies, towards more unusual 3D printing applications driven by innovations possible through additive manufacturing. This forms the foundation for Strategy 3, which then focusses on changing business practice.

PART CONSOLIDATION

The guitar holder designs shown in Figure 4.3 were created using 3D printing because the bespoke guitars they were designed for needed to be displayed at different angles that required the necks and base to be supported. Both products were also 3D printed as consolidated parts to make them foldable and more transportable, without the danger of pieces of an assembly being lost in transit, or difficulties in sourcing the right hardware.

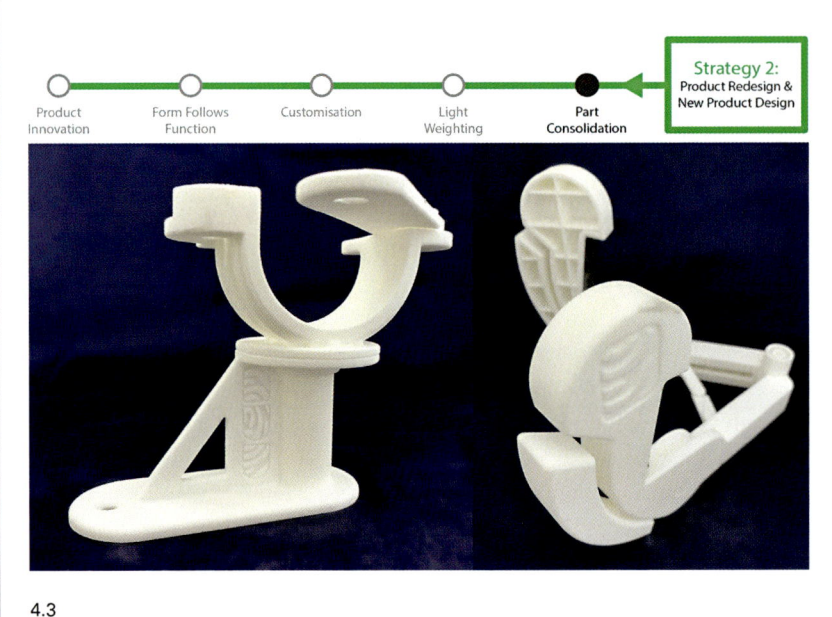

4.3
Example of part consolidation – guitar holder and guitar stand (Diegel).

Faced with the opportunity, and the challenge, to redesign products to exploit additive manufacturing technology for a company, it can be difficult to know where to begin. A good starting point is to look at part consolidation, that is, combining parts that previously existed separately due to manufacturing constraints, or printing an assembly as a single part. This is a relatively straightforward first step to begin exploring additive manufacturing. Without too much redesign, it can streamline production of a part, reduce assembly time, and potentially also remove potential vulnerabilities in the assembly (e.g., joints). Additive manufacturing allows for additional material to be added to weak spots or sections where extra wear is expected. The new design can be run through simulation software to highlight where this should occur [3].

Part consolidation redesign also needs to consider maintenance, particularly where elements may usually need changing out from an assembly as part of a maintenance schedule, and to allow for this process to still happen. Just because multiple parts can be 3D printed together does not mean this is the best decision when looking at the bigger picture. While additive manufacturing is typically more expensive on a part-by-part basis compared to traditional manufacturing, if the number of parts can be reduced, thus cutting assembly time and minimising the need for manual labour, a business case for it can be made.

If needing fewer parts also reduces the chance for part failure – for example, where a part is traditionally assembled with welds – this further strengthens the business case.

Part consolidation provides a good foundation for testing a company's ability to integrate additive manufacturing technology into production practice, as the reality is that even a single part can disrupt production as well as have significant investment requirements for a company. The issues to consider include:

- Selecting the appropriate type of additive manufacturing machine for the parts involved, as well as to fit with the conditions in the factory, and limitations such as space, cost, fume extraction, etc.
- Understanding the material; for example, the size of the powder granules in metal powder bed printing and their relative spherical nature undermines the integrity of the part. Voids created by spaces between the granules, even microscopic, can create weaknesses within a laser-based technology. Sourcing material through the machine supplier, at least to start with, should provide a level of quality control. Take care to check the validity of any warranty or maintenance contract where alternative materials are used.
- Additional equipment will be needed alongside 3D printing hardware – for example, sand blasting for polymer powder-based technologies, and argon gas storage and supply for metal powder bed printers that use titanium. Additional post-processing equipment may be required for highly regulated industries to enhance part quality – for example, hot isostatic pressing (HIP) to reduce porosity of metal prints.
- Whilst a powder bed metal machine may be sold on the basis that different metals can be used within the same machine, the reality is that cleaning out one type of metal powder to replace with another can be problematic. It is advisable to use one machine for one type of powder wherever possible.

All these practical factors need to be considered prior to recommending the introduction of that first part made with additive manufacturing into the production line. Whilst one part may not make a significant impact, it is critical to address the implications of additional parts being integrated in the production line at an early stage before those implications cause disruptions as additive manufacturing production scale increases. This includes a reduction in assembly workers, although some of these could upskill to become technicians for additive manufacturing equipment. It also includes changing the part inventory system and supply chain when fewer parts need to be sourced for each product.

LIGHT-WEIGHTING

The original manifold started as a stainless-steel block. This was then machined/drilled to create input and output ports and channels. Thereafter, holes created during the manufacturing process to allow for the passage of the tool had to be plugged. In this case, the manifold made using conventional techniques weighed 15.2 kg. The new part was designed for 3D printing, with support structures designed into the part. The pipes were angled to reduce the support structures and, wherever possible, additional weight was virtually removed in the 3D CAD model. The holes were shaped to, again, reduce the need for support structures. The final part was made as a single print, required little post-processing, and weighed 1.42 kg – a weight reduction of over 92 percent.

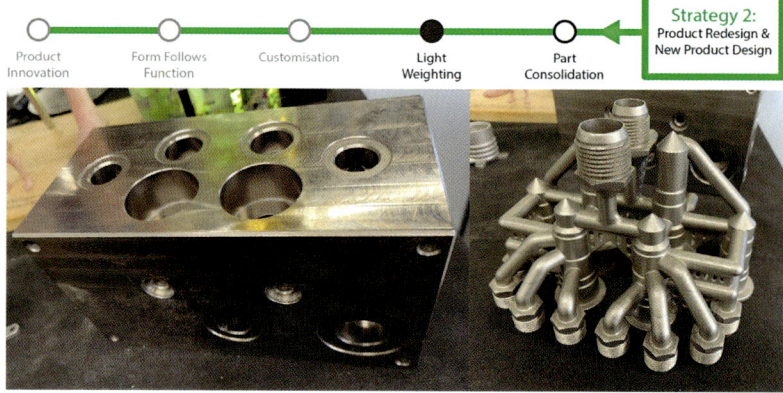

4.4
Example of light-weighting – Atlas Copco Hydraulic Manifold. The original machined manifold is shown on the left, compared with the redesigned and 3D printed version on the right (Diegel).

The next step recommended here, following part consolidation, is focussed on light-weighting (Figure 4.4). Light-weighting is one of the key advantages of additive manufacturing technology. This is because additive manufacturing provides the means to fabricate a relatively idealised version of a part. The constraints of conventional manufacturing, particularly in terms of working with an achievable tool path in both milling and mould part separation, are removed. Organic forms, for example, can be manufactured rather than being constrained to geometric shapes that are more readily achieved with machining.

In essence, the part can be designed more towards the theoretical construct of an optimal product than was previously possible. The design still must work within the constraints of the additive manufacturing selected for the task, but generally, additive manufacturing will allow the designer to fabricate a part that is specified to be far nearer the optimal in terms of strength-to-weight ratio, for example. This means it can be made much closer to the ideal provided in the results of a topology optimisation simulation.

Topology optimisation

Topology optimisation uses finite element analysis in a simulation of load conditions, which provides information on where it is possible to remove material from a part and still achieve performance characteristics based on specified conditions. This may involve specifying where a part is fixed, for example, where it joins another part in the product, and where and to what extent force is applied to different points of the part. Engineers and product designers can fabricate parts more closely aligned to the results of a simulation [4]. Previously, the organic shapes that tend to be the result of topology optimisation were too difficult to make and had to be adjusted to suit conventional manufacturing. With additive manufacturing, design constraints still apply, such as designing for support material, build orientation, and so on, but overall, the relative geometric freedom enabled by the technology means that the results can be much nearer to optimal than previously possible.

This is particularly critical in applications where weight is a major consideration, such as in aerospace [5]. The topology optimised, and 3D printed, Airbus bracket is one of the most high-profile examples of exploring this ability, illustrating the transition engineers and designers have made from their initial limited use of additive manufacturing for topology optimisation in aerospace, to fully embracing the geometries that it enables. Initial interpretation of the results of topology optimisation were essentially a product redesign. In other words, the geometries in the light-weighted product were derived from the original model. However, Airbus went on to evolve the bracket to maximise its performance by designing from the "inside-out," with the performance of the product being the starting point (aka input), with manufacturing (aka outputs) enabled by additive manufacturing, rather than the original product which had been designed for the constraints of traditional manufacturing.

It is important to note here that topology optimisation models are typically rough and still require resolution and design interpretation unless that design had very simple parameters. However good the simulation software, where the results are derived from the original part, the results will not fall outside the existing parameters of that part. In addition, although it will use finite element analysis to advise the designer where material is required, and where it is just excess material, it will not advise the designer on how to design it for manufacturing using additive manufacturing technology. While some software is now beginning to accommodate several additive manufacturing considerations – for example, building orientation and minimising supports – manual intervention by the product designer or engineer is needed to achieve the desired finished product, even when the initial results of optimisation can technically be made using additive manufacturing. Beyond optimisation is generative design, originally the domain of engineering and architecture, but increasingly integrated into product design practice.

Generative design

It is interesting at this point to look at generative design from the same point of view. Where light-weighting takes an existing part and reduces it down, subject to established forces and connection points on the model, generative design works in reverse by creating multiple options for parts that meet the forces and connection points using different geometries. This allows for different solutions that are external to the original part, and therefore, there are more likely to be a wider range of innovative options. However, this still does not provide a solution that is informed by knowledge of designing for additive manufacturing. Therefore, although this can help advise the engineer, the part will still need to be designed for the additive manufacturing technology being used. Nevertheless, software is still a way off from being able to make different generative designs, or even topology optimisation, based on the specific 3D print technology being employed.

Just as with specifying for other conventional manufacturing technologies, such as injection moulding or CNC routing, there are constraints. It is not possible to print just anything, and even if it were, there would be no benefit in doing so. The reason the two activities are linked in people's minds is because without additive manufacturing, it would not be possible to manufacture the organic complex parts that are often the result of topology optimisation. However, this does not mean that the part is optimised for manufacturing using 3D printing. This is a common misconception, and to use the software only is a superficial and frequently inappropriate solution.

In both cases, though, the key step is for the product designer to use the generated part for optimisation simulation to inform the design. It is not to rely on the output to be suitable, or maximised, for additive manufacturing. This will require further steps. The design rules for additive manufacturing described later in the book still apply. It will need refining to the constraints of that technology, such as the need for incorporating support structures – their location is a critical element in DfAM. Most of the options provided by generative design or even topology optimisation will not meet appropriate detailing criteria without further work. Detailing for the correct additive manufacturing technology, and for the task, will still need to be incorporated and, preferably, drive design considerations. In addition, as with all designs, the part must connect to other parts of the assembly as appropriate, function effectively without the additional material removed in the process, and be able to be maintained.

DESIGNER'S COMMENT

One of the keys to design for AM is getting rid of material that doesn't provide real value-adding functionality to your product.

Cutting-edge optimisation software is developing, by which the part is generated in line with the constraints faced in real time, thus responding to the forces as the part "grows." This is an interesting generative design approach, as it is directly intended for use with additive manufacturing. "Live parts" was developed by the MIT team who developed Desktop Metal, a company which is leading innovation in additive manufacturing. However, this is still in its early stages of development at the time of writing. There are product designers who are using generative design in a more creative way, and who are working specifically with additive manufacturing, their work informed by expertise in the technology. However, much like conventional manufacturing, testing would still need to be planned for a company before additive manufacturing a generatively designed product. This testing would allow the identification of weak points, flex, compression, shear stresses, and tension, though the logic for such testing will be different to conventional manufacturing.

Material testing would be different, and consistency between prints or different machines of the same technology would need to be assessed. New testing procedures would need to be developed to meet the functional requirements of parts, particularly if there is no precedent of testing or body of knowledge for the product designer and engineer to draw on for the application.

Lattice structures

Arguably one of the most significant characteristics possible with additive manufacturing is produced through light-weighting based on removal of material in the internal structure of the form itself. Think of a part as constructed from a cellular foam shaped into an object, with the voids in the foam itself varied to create different performance characteristics within a single part. A more controlled version of this approach for product designers is the use of lattice structures. Different forms or scale of lattice can be used within a single part to create similar variations in performance.

The ability to design and manufacture lattice structures within a part, either the body or the walls, offers significant potential for the product designer in the redesign of products. It opens possibilities in design and construction that were previously prohibitively difficult. Lattice structures can be used with most additive manufacturing technologies, with structures that can be adapted not only between parts, but within parts themselves [6, 7]. For the product designer interested in becoming a specialist in additive manufacturing, designing lattice structures from the ground up, and understanding their use, is a key skill. It is also one of the identifying traits of a 3D printed product compared to any other manufacturing process.

For product designers for whom working with additive manufacturing is a relatively minor part of their portfolio, simple tools to create lattice structures are available in accessible software that provide a good starting point for integrating the practice into designs. Free software programs, such as Autodesk Meshmixer, are readily available. These programs enable the conversion of solids to lattice with a range of different structures to select.

More advanced CAD software, such as Materialise 3-Matic or nTopology, increasingly provides tools for the development of multiple customised lattice structures to be specified within a single part, thereby reducing solids to light-weight structures throughout, in a predetermined pattern. This approach provides opportunities to vary the properties of the lattice in different regions to affect functional properties. Imagine a part with flex structurally built into one end, and rigid at the other, because of changes to the internal geometries of the lattices. Think of a product not as an assembly of parts, but as a single form with the equivalent of different material performance characteristics within the same part constructed from a single material. Lattice structures allow the product designer to realise product outcomes that react to forces differently within the same part, although made of the same material, and to control those differences to a high degree of accuracy both in terms of performance and location. However, these tools require advanced skills to master and are ideal for professional product designers.

The design of lattice structures within a part needs to follow the same design rules for additive manufacturing as those that apply to larger components. Additive technologies that do not require supports – for example, powder bed fusion processes – are ideal for producing lattice geometries, providing the most freedom. However, they can be produced with any of the additive technologies, with the angle of the components of each lattice being a key consideration during design and fabrication to minimise the need for support structures because these would be difficult to remove within a lattice structure. As a light-weighting technique, lattice structures are one of the most recognisable qualities of additive manufactured parts that are still in their infancy. Software developments and more advanced simulation capabilities could improve the integration of this approach into a design workflow.

CUSTOMISATION

The lugs shown in Figure 4.5. were built as a parametric system to customise the geometry to suit a specific patient. These were further customised through the addition of graphics and 3D printed on a HP Jet

Fusion 580 full-colour machine in polyamide 12 (nylon) (designed by James Novak). The bottom row shows some of the process of FEA analysis on an original lug block model, topology optimisation, mesh smoothing and conversion to surfaces, and final FEA analysis used by Nalinda Dissanayaka on the development work for this lugs project.

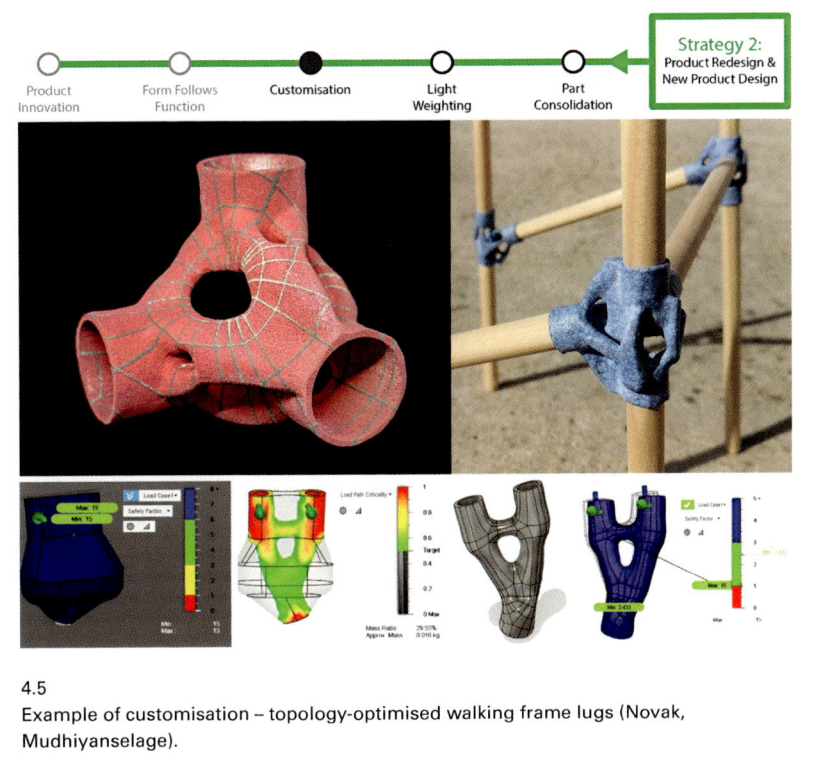

4.5
Example of customisation – topology-optimised walking frame lugs (Novak, Mudhiyanselage).

3D printing is particularly useful when working with products designed for children because of the variation in the sizes needed. This is particularly true for walking frames, with a range of anthropometric and ergonomic requirements, combined with loading conditions applied to the frame during use. In this example, measurements and data from a patient can be entered into a parametric CAD system which then generates the block geometries for the lugs. These are then topology optimised to remove excess material, resulting in a one-off walking frame design without the need for numerous adjustable components or attachments. The child can also customise the lugs to make them appear less medical, applying images or colours where they like. Dr Novak developed this range of lugs for children's walking frames, with Nalinda Dissanayaka investigating through simulations and practical testing the performance characteristics of different forms of the lugs.

The ability for individuals to customise products can add value to a product range. However, this approach works as part of a product service system rather than as a stand-alone product offering. The platform for interaction, online or otherwise, and workflow to service that customer interaction must also be designed into the production and supply system. This additional factor can counteract the value added by the service, not only in terms of the costs involved but also in terms of the brand image and value, as any problems may damage the brand value. Thus, it is vital to evaluate if this strategy is a viable new way of working for the company.

For Strategy 2, within this staged approach to technology adoption, the recommendation is to limit customer interaction to customisation only, over the more involved practice of product personalisation (discussed in Strategy 3 as part of rethinking business practice). Personalisation involves working with individual data provided by the customer. Customisation is less involved [8], and so one should think of it as providing a degree a choice for the customer, but within constraints – for example, the ability to add a name to a part within a product or to choose from a set of different textures for a 3D printed part.

Customisation should be included at this stage because it prepares the workforce, and company operations, for the ability to work with different parameters within a single product while simultaneously avoiding the radical rethinking of operations, particularly in terms of online platforms that personalisation requires. For this stage, a company can start to build an interactive online presence if it chooses to do so, with options for the customer that are selected from a limited library of resolved parts only. These parts can be predesigned, tested, and priced in advance to ensure that each part can be validated and manufactured before its inclusion in the end product. In contrast, personalisation is much less predictable. This customisation approach builds knowledge across the company on design for 3D printing whilst keeping the range of products being designed within known boundaries. It may also be possible to integrate traditionally made parts at this stage with a mix of 3D printed ones, or to make the parts using conventional technology but also 3D print them as part of a marketing tool to test the waters before a full-blown shift to Strategy 3. Either way, keeping the range of parts limited to options that do not push the boundaries too far can be a good starting point for building confidence in the additive manufacturing technology selected for the task.

Positioned between conventional customisation and personalisation that conforms to unique data provided by the customer, there are innovative examples of product service systems built on an interactive digital interface that exploit the characteristics of additive manufacturing technology. Fung Kwok Pan, a Singaporean UI/UX designer, experimented with the possibilities of 3D printing through his work called Fluid Vase, featured in the design magazine *Designboom*. The vase form simulates moving fluid, and the customer can pause the animation whenever they choose. The result is then 3D printed. This work is experimental, very much at the interface between art and design, but it illustrates the potential of 3D printing to allow a new spectrum of geometries and new ways of co-creating.

One of the most sophisticated online platforms currently available to customers to engage with, and order a customised 3D printed product from, is supplied by Nervous System (https://n-e-r-v-o-u-s.com/). The interface provided by this company arguably provides a model of how all products could be supplied in a digitally enabled future. Their kinematics work illustrates how digital technology can be used to rethink product design and fabrication. They use algorithmic design based on parametric principles to enable their 3D models to be broken down to code suitable for embedding in a web platform. Overlaying this is an interactive interface that allows anyone to easily engage with a model of a piece of jewellery or fashion and alter its geometry or features to create a unique solution.

This ability to "push and pull" the geometry onscreen is viable only because of the ability to 3D print the outcome; otherwise, it would simply be a novel virtual experience. It is important to note that the result is not personalised as such, there are constraints built into the system, and the customer must work within the limits the algorithm provides. However, there is a degree of customisation which remains beyond the norm. This type of sophisticated parametric model is arguably the basis for product design of the future. The interactive nature of the interface adds to the sense of co-creation for the customer. The designer becomes a creator of systems rather than singular products, which allows for almost limitless variations.

FORM FOLLOWS FUNCTION

Geometrical complexity is possible for functional as well as aesthetic reasons. Exemplars on using the capabilities of 3D printing for creating value-added products demonstrate a form-follows-function approach. The Canal House shown in Figure 4.6, by DUS Architects in Amsterdam, explores how the actual fabric of a building can be designed from the inside out to respond to place, environment, and operation of the building [9]. For example, the thermal mass of a building could be varied throughout the structure, and so too the control of the flow of fluids through the building. The performance of structural geometries, shown in Figures 4.6 and 4.7, by DUS are further adapted by the addition of fillers into the spaces created using 3D printing. Whilst many architects use a robot arm for 3D printing, DUS used a very large FFF machine, much like an oversized desktop.

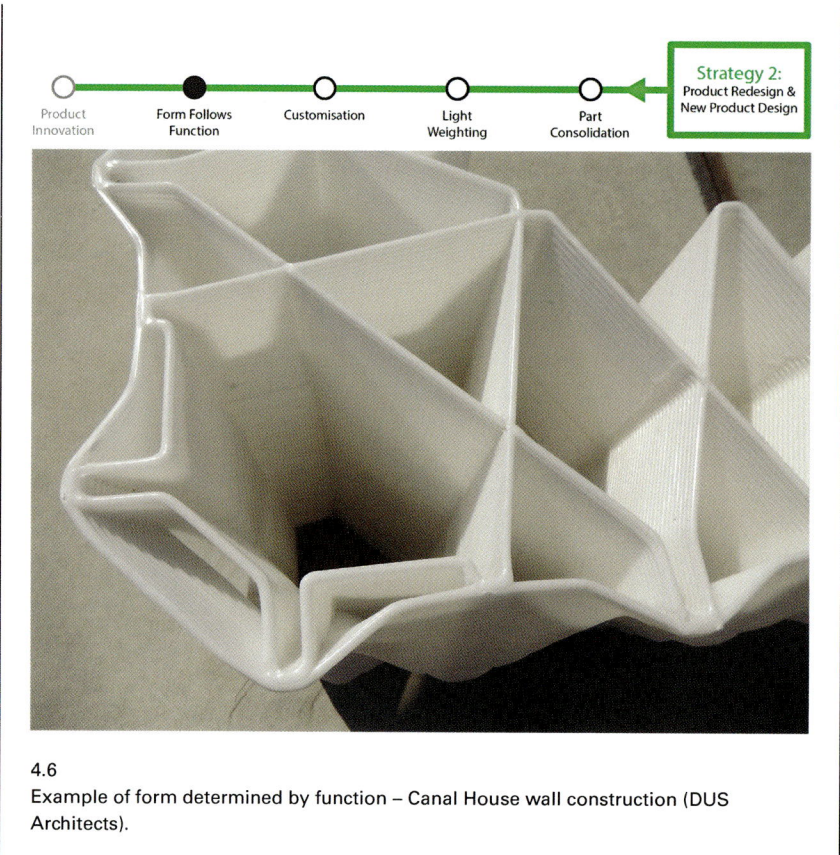

4.6
Example of form determined by function – Canal House wall construction (DUS Architects).

The architect Louis Sullivan, known for steel frame constructions, is credited with the use of the term "form follows function." Sullivan was referring to the complexities of nature in his comment, but in product design, this phrase became synonymous with Modernism, with a focus on minimalism and the removal of decorative elements that performed no function. Dieter Rams followed up on this principle with the statement "less but better" during the functionalist industrial design approach dominant in the mid-twentieth century. Until the advent of movements such as post-modernism, and experimental designers such as Italian designer Ettore Sottsass, products with an austere functionalism dominated product design culture and teaching.

It is interesting to consider how the recognisable characteristics of "form follows function" from the last century, such as geometric shapes and clean lines, would be subverted by additive manufacturing. Additive manufacturing allows for design "from the inside out." In this, it has more in common with nature than human-made forms that other manufacturing technologies are optimised to produce. Bone structures, as explored by designer Ross Lovegrove, for example, or the complex geometries of MIT researcher Neri Oxman's Material

4.7
DUS Architects 3D printed formwork creates voids that can then be filled with different materials, such as concrete or pulp.

Ecology [10], bear little resemblance to the austere simplicity of the work that dominated the mid-twentieth century and shaped industrial design teaching as it emerged in tertiary education. Yet, additive manufacturing enables a form-follows-function approach that transcends products made using conventional manufacturing, and the results are startling when viewed both functionally and aesthetically.

A good example of this exploration of geometrical complexity, driven by functional and performance considerations, is the use of additive manufacturing for the design and manufacture of high-performance heat exchangers (see the technical examples section at the end of Chapter 6). Instead of analysing existing parts for potential improvements and then seeing if the results could be 3D printed, designing with, rather than a solely technical design for, additive manufacturing results in products that are designed to maximise the functional outputs, based on an understanding that the technology itself is integrated into that part rather than as an add-on in a linear process.

Developing the design during this approach starts with the fundamentals of what the part is required to do at a much earlier time than would normally be seen with conventional manufacturing techniques. With additive manufacturing, it is then possible to work outwards to map the ideal geometries for that functional requirement, and then expand the structure out, exploiting the capabilities of the technology. Any reduction in weight is measurable and potentially a benefit to the client, whereas other potential benefits are less easily quantified. For example, where multiple parts have been consolidated into one, this could lead to a reduction in potential weak spots where components are joined. This saves the company money and preserves its reputation over time, as well as the assembly costs of joining those parts as described in the part consolidation section earlier in this chapter. Understanding the complex and interrelated impacts of the differences created with additive manufacturing is key to maximising product designed with the technology.

This stage of technology adoption can be driven by structural analysis and simulations. However, the difference between light-weighting and this approach is that through this approach, the product is designed from the starting point of the function, and built outwards, rather than from an existing product that is to be adapted using fabrication enabled by additive manufacturing. This approach is more difficult to visualise, but essentially it is based on working with a theoretical construct that is then 3D printed, rather than a product that is optimised.

DESIGNER'S COMMENT

Most parts are designed in a particular manner and with certain materials for historical reasons rather than what might be most appropriate were the product to be designed from scratch now. Sometimes it is difficult to rethink a product rather than just redesign it. A starting point is to think not of what material a part should be made of in a replacement product, but rather of what function it should have, and to work outwards from there.

PRODUCT INNOVATION

In Tuber9, in Figure 4.8, Dean demonstrates how to exploit the opportunities of 3D printing, rather than working against the change in practice it enables. Dean's early work was built within the gaming software Virtools. However, a Rhino plug-in called Grasshopper allowed him to use software better suited to design in that it allowed a suitable 3D CAD model to be built within the software, or to be imported and manipulated. Tuber9 was acquired by the Museum of Modern Art in New York in 2006. The lamps created in this project were displayed alongside the animation used to create them. To create the animation, a "random generator" algorithm, which selected numbers from a specified range, was linked to a base computer model of the lamp. This allowed the model to be extended "randomly" but within a constrained space.

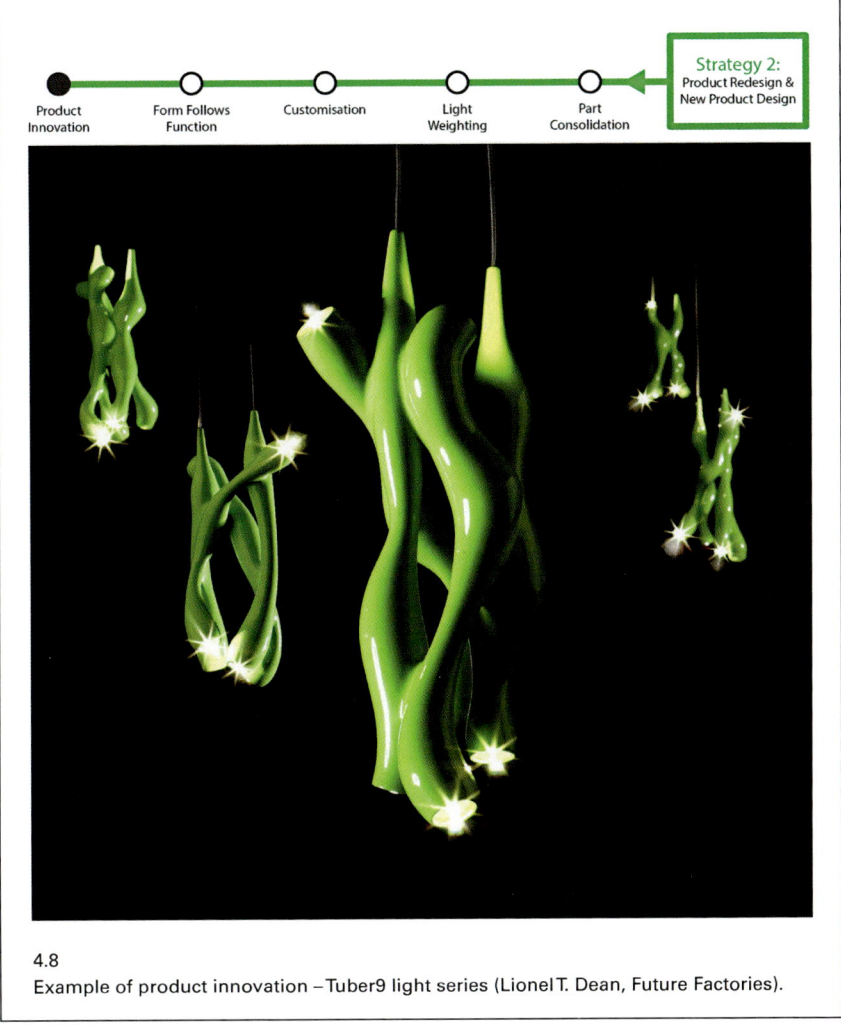

Product Innovation — Form Follows Function — Customisation — Light Weighting — Part Consolidation

Strategy 2:
Product Redesign & New Product Design

4.8
Example of product innovation – Tuber9 light series (Lionel T. Dean, Future Factories).

Arguably, the most exciting aspects of working with additive manufacturing are situated within the product innovation space. In this section, the discussion focusses on the potential of additive manufacturing to enable innovative product design, and how it can be used to meet that potential. This is the combination of all sections covered in Strategy 2 so far – in other words, part consolidation, light-weighting, customisation, and form follows function. However, the discussion will avoid, as far as possible, straying too far into business innovation, a topic covered in Strategy 3. Instead, the focus will be on the design of innovative products enabled by additive manufacturing. The underlying approach, then, is to design a product that could not be made any other way and that demonstrates to a company, as an exemplar, what additive manufacturing could offer in terms of innovative product design for the long-term.

> **DESIGNER'S COMMENT**
>
> The more geometrically complex a part is, the better suited it is to 3D printing. But the converse is also true! If a part is very simple, there may be better ways to make it!

As the culmination of approaches covered in Strategy 2, product innovation is not ideal as a starting point for established companies due to the level of disruption it would create to implement simultaneously. However, a start-up or research and development (R&D) group may choose to go all out and begin here, guided by experienced DfAM product design expertise. The goal of this may be to leapfrog the competition, or to radically challenge the status quo, which in some circumstances may be the only way to facilitate organisational change rather than the staged approach of the strategies in this book.

The reason product innovation is the final approach in Strategy 2 is that it requires a complete rethinking of practice for a product designer trained in conventional design for manufacturing and is, therefore, a good basis for a change of mindset in design using this technology. Conventional design for manufacturing involves developing components with the restrictions created by fabrication methods that rely on tooling. Components are assembled into a product. This is not necessarily the case in additive manufacturing, as complex geometries can be 3D printed in a single part, or as an assembly 3D printed as a single part, or even with captured parts within a print. Additive manufacturing allows for these "geometrical impossibilities" in relation to conventional manufacturing to be combined and facilitate radically new products, including new systems to engage with consumers. Therefore, in value adding with additive manufacturing, one of the starting points whilst learning to work with this technology is to explore with test pieces that demonstrate that capability.

If a company has worked through several of the Strategy 1 and Strategy 2 approaches, they will have already developed test pieces alongside upskilling their workforce, identifying the most suitable products and markets to address. They would have already invested in 3D printing hardware and associated software, or partnered with a supplier who facilitates this, and would be planning their additive manufacturing strategy for the next five to ten years. However, in bringing it all together, new test pieces are needed that combine the work done already or apply the knowledge gained to a completely new venture. The challenge is to decide where to start and what would demonstrate appropriate and complex geometry, even where function is taken out of the equation. An alternative is to select a product to work on that is basic in terms of functionality but allows for geometrical complexity in the form, such as a light shade. The intent is to allow a product designer freedom to begin test pieces that move their thinking from standard geometries into more organic shapes, and to be able to see the results in a physical form.

Successful product innovation, as framed by Strategy 2, is enhanced by collaboration. Engineers and product designers can work independently on additive manufacturing, creating solutions to different problems. Combining the two points of view can enhance the opportunity for innovation. However, the current gulf between disciplines with an interest in additive manufacturing, in terms of ambition, ways of working, skills, and body of background knowledge, is significant. Even communication between the disciplines is challenging, as there are language differences and different values in terms of what is produced and why, particularly through the dominance of engineering and medicine shaping the evolution of the technology. The interest levels in the product design disciplines have been much slower to develop. Yet, product designers have the creative mindset needed for innovation, and the ability to work with uncertainty and risk, more than other technical fields that may leverage some of the freedoms of additive manufacturing but operate within tightly regulated industries that constrain creativity. This is a valuable skill in terms of the development of the industry.

WHAT MIGHT HAVE HAPPENED WITH THE FICTIONAL SME …

Fox Coral Solutions' first foray into commercial products using 3D printing went well. They discovered they could create value-added detailing products to their motorbikes in the after-market range. Thanks to some of their workers, who had already been experimenting with small 3D printed add-ons for their personal rides (as in Figure 4.9), they had been able to use existing desktop printers to quickly prototype ideas and gain feedback at weekend track days. This built confidence in the product opportunities, while Meeks from LND gave Fox Coral Solutions the confidence to invest in an SLS machine. Additionally, they also enthusiastically adopted dip-dying techniques to add additional customisation options.

4.9
Fox Coral Solutions can now supply their traditional after-market motorcycle parts like foot pegs (green), while also providing customised 3D printed inserts (orange) that can be sold as part of the package, or individually curated by riders online.

Fox Coral Solutions were also open to using initial test prints from the SLS machine to enhance their knowledge in how to adapt their traditional automotive parts post-production techniques to apply a high-end gloss finish to 3D printed parts. Their plan is to train key workers in using metal printing, sending them across the country to an upcoming five-day intensive workshop. In the next 12 months, they planned to invest in SLM so that they can include more functional components for their performance customers. These included some light-weighting of brake callipers and brackets. Although they intend to start with a small build size to keep costs manageable, they are already discussing expansion plans as the team looks at opportunities for increased performance, lower maintenance, as well as stylistic features.

DESIGNER'S COMMENT

When designing something new using AM for series production, we can't just design for function; we need to take control over the production of the parts too. This means designing with the characteristics of the additive manufacturing process (both positive and negative) fully in mind. We need to design the build at the same time as we design the part.

On to the final stage of the journey …

REFERENCES

1. Ding S, Ng BF. Particle Emission Levels in the User Operating Environment of Powder, Ink and Filament-based 3D Printers. *Rapid Prototyping Journal* 2021;27(6):1124–1132.
2. Floyd EL, Wang J, Regens JL. Fume Emissions from a Low-cost 3-D Printer with Various Filaments. *Journal of Occupational and Environmental Hygiene* 2017;14(7):523–533.
3. Schmelzle J, Kline EV, Dickman CJ, Reutzel EW, Jones G, Simpson TW. (Re) Designing for Part Consolidation: Understanding the Challenges of Metal Additive Manufacturing. *Journal of Mechanical Design* 2015;137(11).
4. Barbieri L, Calzone F, Muzzupappa M. Form and Function: Functional Optimization and Additive Manufacturing. In: Cavas-Martínez F, Eynard B, Fernández Cañavate F, Fernández-Pacheco D, Morer P, Nigrelli V, editors. *Advances on Mechanics, Design Engineering and Manufacturing II*. Lecture Notes in Mechanical Engineering: Springer; 2019. pp. 649–658.
5. Meng L, Zhang W, Quan D, Shi G, Tang L, Hou Y, et al. From Topology Optimization Design to Additive Manufacturing: Today's Success and Tomorrow's Roadmap. *Archives of Computational Methods in Engineering* 2020;27(3):805–830.
6. Feng J, Fu J, Lin Z, Shang C, Li B. A Review of the Design Methods of Complex Topology Structures for 3D Printing. *Visual Computing for Industry, Biomedicine, and Art* 2018;1(1):5.

7. Reddy AH, Davuluri S, Boyina D, editors. 3D Printed Lattice Structures: A Brief Review. *2020 IEEE 10th International Conference Nanomaterials: Applications & Properties (NAP)*, 9–13 Nov. 2020; Sumy, Ukraine: IEEE.

8. Hu SJ. Evolving Paradigms of Manufacturing: From Mass Production to Mass Customization and Personalization. *Procedia CIRP* 2013;7:3–8.

9. van Baarsen SB, Schonwalder J, Houtman R, van der Veen AC, Vermeulen H, de Haan S. The 3D Printed Canal House. *Proceedings of IASS Annual Symposia, IASS 2015 Amsterdam Symposium: Future Visions – Emerging Technologies: 3D Printing and Robotics*; 2015 17–20 Aug. 2015; Amsterdam, Netherlands: IASS.

10. Antonelli P. *The Neri Oxman Material Ecology Catalogue*. New York: The Museum of Modern Art; 2020.

5 Strategy 3

Digital business innovation

This is arguably the most sophisticated strategy, and the hardest to define, as it involves anticipating, identifying, and addressing what could be the unanticipated consequences of disruptive change. This approach is also the most complex and cross-disciplinary. It is based on a company making significant changes to business practice to maximise the opportunities provided by additive manufacturing. The strategy is presented as a series of successive approaches that build from least disruptive to most, each with the potential for improvements and pitfalls. This strategy is also relevant to entrepreneurs and start-ups whose agility allows them to begin working at the cutting edge of additive manufacturing unencumbered by traditional manufacturing paradigms and, potentially, without the preliminary investigations identified through Strategies 1 and 2.

5.1
Overview of Strategy 3 approaches to 3D printing.

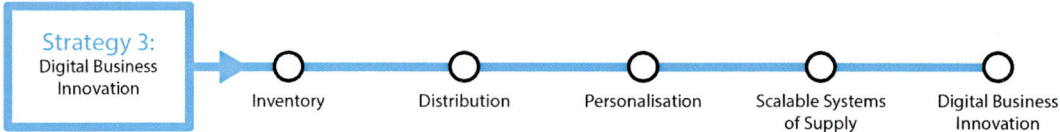

Strategy 3:
Digital Business Innovation

Inventory   Distribution   Personalisation   Scalable Systems of Supply   Digital Business Innovation

THE MULTINATIONAL: AN AMBITIOUS, INTERNATIONAL COMPANY LOOKING TO BE A WORLD LEADER

The board of the fictional firm, Hydnora Technologies, is led by Berriz Petrus. He took over from Sariba Tumelo when she moved into an industry advisory organisation. The board includes the heads of departments and external advisors on business growth and administration. After a year, Sariba contacts Berriz and advises him to investigate how to expand the company overseas. She says that new grants have been recently announced to support the scaling up of mid-level companies with an emphasis on business innovation which is driven by digital technology, and in particular digital convergence. Berriz reports back to the board, and they create a rapid response research team to understand the opportunities and the challenges involved (Figures 5.1 and 5.2).

Glenn Easterman presents the findings to the board. He says the team studied the company's current capability and potential futures. Their advice is

DOI: 10.4324/9781003122203-6

5.2

to look at digital business innovation enabled by digital technologies. No one on the board is sure what that means, and Glenn admits he is not sure either, but he thinks they should maximise the advantages they have – the company is already a leader in producing lightweight parts for a range of industries using 3D printing. He suggests this may be a good starting point to consider their options, building on their market advantage of established 3D printing exper-tise. The board tasks the team with finding a design consultancy specialising in additive manufacturing that has a good understanding of global develop-ment in digital technology and business innovation. The team contact LND Consulting.

For Droov, this is the perfect opportunity to take working with additive man-ufacturing in the company to the next level, in business terms. The company is just the right size and has the right level of expertise, ambition, and foresight to be able to consider a step-change in the way they work and what they do, rather than an incremental change in practice. He outlines his thoughts to the board and the research team are enthusiastic, though some members of the board, most notably Claire Torres, are less enthusiastic. Claire finds it difficult to see how this could work without undermining the market position the company already has and worries it may not be worth the risk. Droov explains there is a step-by-step strategy, even for the introduction of a paradigm shift, however much of a contradiction that sounds. Claire votes with the rest of the board to see how it could work for them. Droov outlines Strategy 3, and how he sees the company's future.

WHY WORK WITH STRATEGY 3?

Additive manufacturing is a disruptive technology. It is aligned with the range of digital technologies that have emerged over the last three decades. These include digital communication tools (e.g., mobile phones and the Internet), data monitoring, data analysis, and machine learning, tools that are impacting not only consumer behaviour but societal priorities, production practices, distribution, and lifecycle. The practicalities of additive manufacturing as fabrication technologies have developed considerably over the last decade. The underlying profile of the technology has finally shifted from a prototyping technology into a production technology. Many companies that began with individual machines for prototyping are moving towards manufacturing and promoting production-scale facilities for companies in many different fields. Desktop Metal is a good example, as their starting point was providing a studio system to create high-end, one-off parts, mostly for prototyping; but their ambition was to create a production system for the additive manufacturing of end-use parts in large volumes. The size of the printers illustrates this, from a desktop model to a factory model.

This shift in thinking over the last decade changes the role of the product designer. More than ever before, product designers need to research complex challenges, interact with stakeholders, understand supply chains and product lifecycles, and respond to increasingly "wicked" problems [1]. Technology platforms enable co-creation and require a ramping up of digital skills for product designers. Further, business practice innovation is also becoming an equally critical skill. The type of product, and its individuality, means that designing for it involves a change in thinking about business practice as much as design for process. Yet, many manufacturers currently underestimate the changes that the technology can provide. Through the previous strategies of this book, additive manufacturing can be subsumed into conventional production as simply another manufacturing facility if desired, minimising disruption while building awareness and skills, and testing market opportunities. The broader potential for digital business innovation can easily be overlooked through this process.

DESIGNER'S COMMENT

For entrepreneurship to succeed, you need to be able to fail fast and fail often. With additive manufacturing, you can fail extra fast and extra often.

This chapter identifies several areas where product designers can add value to the additive manufacturing adoption journey of a company that is ready for the step-change required, as identified in Hydnora by Glenn and Droov. Where Strategy 1 was about early-stage adoption with minimal disruption, and Strategy 2 was about new product innovation, Strategy 3 embraces disruption and the new ways of conducting business enabled by the digital nature of additive manufacturing technology. While the concepts are likely to be on a company's radar after beginning their additive manufacturing journey – for example, digital inventory and new distribution models will be a natural extension of manufacturing any end-use application 3D printed parts – the goal is now to embrace these new opportunities that come with producing parts directly from digital information. This does require a willingness to step back from existing business practices and rethink what is possible. It also requires a good understanding of additive manufacturing, and preferably of the associated digital technologies, such as 3D scanning and creating an interactive, online customer interface.

For a company such as Hydnora, big picture thinking is in their DNA, and they are looking to take a leadership role in their industry in using the technology. The more they find out about the potential for significant savings, supply chain security, and added value through reorganising elements of their business model, the more committed Berriz becomes. Even Claire is cautiously – though nervously – optimistic.

DIGITAL INVENTORY

The example shown in Figure 5.3 was a collaboration between Griffith University, the Humanitarian Innovation Fund, RedR (Australia), and HK Logistics with Oxfam G.B. [2]. The work analysed the inventory of the Water, Sanitation and Hygiene (WASH) project, East Africa for parts that could be viably redesigned for 3D printing, with appropriate digital files developed and validated. While traditional supply chains provided mass-produced parts like connectors and adaptors for water pipes, operators on the ground found that specific pieces they needed were sometimes unavailable. This may be because of supply issues, or the part was damaged in transit, or the part geometry needed may have been outside the scope of standard supplies available.

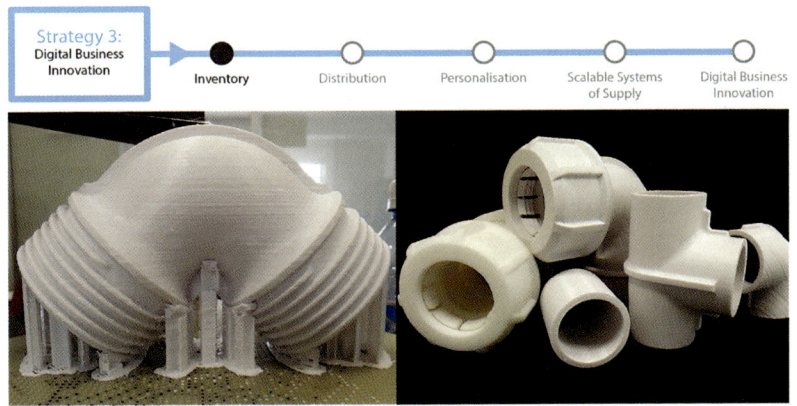

5.3
Example of a digital inventory project – parts digitised, redesigned, and validated in Australia for 3D printing, then 3D printed in Nairobi, for the Oxfam G.B. WASH project (Loy, Tatham, Healey, & Tapper).

Where parts were critical for the flow of water, a replacement had to be found. With conventional supply chains taking approximately 12 weeks to reach Nairobi, temporary replacements were usually cobbled together out of available parts in the local market. However, these were described as unsatisfactory and unreliable by the field operators. 3D printing provided an alternative approach, enabling the right part to be 3D printed rapidly and locally until normal supplies became available. This also opened opportunities for digital networks to support what was happening on the ground. Parts were able to be designed and/or validated in other locations and shared via the Internet.

One of the first steps in introducing new business models enabled by digital technology disruptions is to consider the potential to create a digital inventory for the company. This strategy helps provide a baseline for understanding the digital potential of a company based on its existing inventory, identifying lead parts or products than can be digitised, and then using this as the basis for digital business innovation. A good starting point is to look at the spare parts a company has, especially those it must keep for seven years or so after it has stopped producing a product (often mandated by law). This inventory ties up warehouse space and, therefore, money. Parts can be damaged or lost during storage. Supply may also run out early if demand is higher than anticipated, and it can be extremely difficult and expensive to remake parts many years down the track for a product that is no longer in mainstream supply.

Warehouses that hold unsold and unused inventory consume electricity for heating, cooling, and lighting. Replacing physical inventory with digital inventory would green the supply chain. Physical inventory not only needs to be transported, but it also takes up a lot of shelf space while it waits. In contrast, a digital inventory – or design files for a 3D printed machine part – is cheap and easy to store and transport. 3D printing technologies could help clean up the manufacturing process if their unique capabilities are used [3]. Any parts that can be kept as digital inventory, and then printed on demand, will release space taken up by inventory. Whilst the prints themselves may be more expensive than producing the original as a batch, the benefits of a virtual inventory may outweigh the printing costs. The challenge of this, as noted in the Bridge Manufacturing section of Strategy 1 (Chapter 3), is that the spare parts were likely designed for traditional manufacturing methods, and direct 3D printing of these may not result in parts that perform as desired. This may be corrected by using a different material or may require design modifications based on an understanding of the capabilities of a particular 3D print technology. However, the time investment to redesign parts and, therefore, the cost to do this must be weighed against the other options available.

From this starting point, the next step could be to identify parts in a production that are critical for a company's operations. Think of parts for a conveyor belt operation, or specific types of hardware. If without these a company's operations are disrupted, then they are worth recording as digital inventory in case of disruptions to the supply chain, as happened during the start of the COVID-19 pandemic. For these parts, intellectual property (IP) issues can apply, which may need to be worked out with the original supplier. However, it is also opportune to rethink and redesign the part specifically for additive manufacturing, with a view to printing it on a specific technology locally available. This will likely change the part to such an extent that it mitigates IP issues. It may also be an opportunity to identify why that part is critical or vulnerable, and perhaps redesign an assembly as a part consolidation, as described in Strategy 2, or add material at the point of weakness during its translation to a 3D printed part.

The original equipment manufacturer (OEM) of a part or product has IP rights and, frequently, a system in place where maintenance contracts are invalidated by changing out original parts. In addition, storing and sending digital files arguably creates a security vulnerability in the supply chain. However, where parts are redesigned for 3D printing, as a spare part or unique part, new IP will be created. Working with OEMs, companies could design new parts for purpose. Equally, it may be time to rethink the relationship, allowing for the licencing of production on site where suitable.

Aligned with the development, digital storage and production of inventory, internal distribution of this inventory also enables new business practice. Digital product models can be sent worldwide for 3D printing on site, or at least in local proximity to the site where they are needed. Data can be collected for a bespoke item at the source, then sent digitally to the company headquarters, designed, and tested, then digitally sent to a service bureau near to the site or even the customer. Depending on the size and resources of the company, the 3D printing facilities may even be available in a branch of the same company, sent to be printed in-house on the alternative site. There are challenges involved with this approach that product designers can help resolve. Quality control with calibration tools, for example, need to be developed alongside the part to ensure that even when it is printed externally, it matches the original validation tests that may have been performed at the headquarters. In addition, materials may be from a different source which could affect fabrication and, therefore, must be taken into consideration.

DISTRIBUTION

During the early stages of the COVID-19 pandemic, the rapid shutdown of global supply chains led to a scarcity of many essential medical devices and personal protective equipment (PPE). However, the widespread availability of 3D printers in businesses, libraries, schools, and homes meant that new methods of community-led production were possible, and product designers, as well as engineers, clinicians, makers, and students, were able to design, test, iterate, and share new forms of PPE via the Internet [4]. An example is shown in Figure 5.4. The distribution of products changed from centralised forms of production to decentralised and distributed methods. This did raise some concerns over the safety and validity of what was being uploaded to the Internet, particularly in the first months of the pandemic [5, 6]. However, ultimately it showcased the opportunities for online distribution of files for additive manufacturing on a global scale.

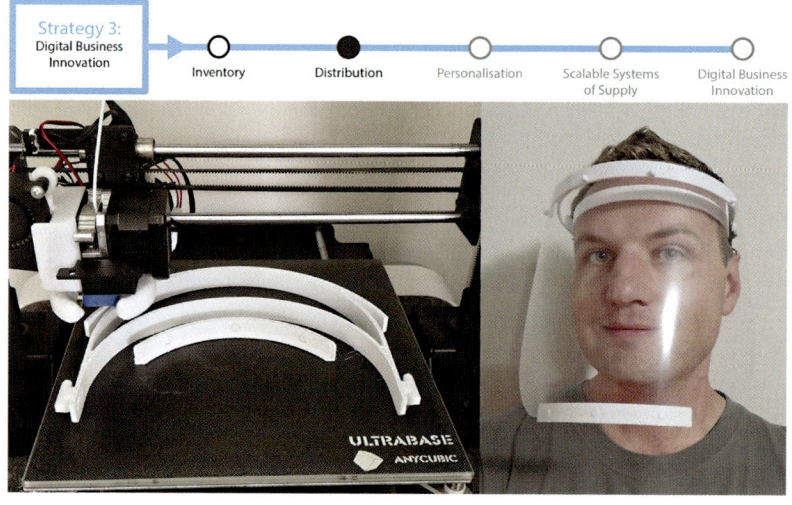

5.4
Example of a distributed manufacturing context – low-cost, open-source 3D printed face shield called the RC3, designed by Prusa Research (Novak).

Once digital inventory has been created, the next step for a company is to look at how the virtual inventory can enable changes of practice beyond the confines of the company or factory. Digital inventory saves on warehousing, but it can also potentially save on distribution costs. On-demand manufacturing – in other words, parts that can be manufactured as, when, and where needed, rather than having to keep a large stock of parts on-hand and transport them – contracts the supply chain and reduces the environmental footprint of the parts. This is often known as Just-in-Time (JIT) manufacturing.

For a company that supplies customers at some distance from their facility, even overseas, it may be possible to send the digital file to a facility nearer the customer for 3D printing there. As a spare part, this approach is more likely to be viable compared to sending actual products this way. This owes to the potentially high costs of fabrication. Managing that fabrication off-site can be offset by the reduced cost of shipping an individual part to a one-off location. However, there are quality control issues that must be considered and managed, as noted in the previous section on inventory. If the performance of the part is dependent on critical function or fit requirements, then the quality of the 3D print needs to be agreed and monitored, and the outcome validated. This is not as straightforward as one might assume.

At this time in the United States, research into the validation of metal parts for aerospace, for example, is limited to individual printers. Changing printers can invalidate parts being produced in certain circumstances. Even where this standard of validation is not required, the reality is that the settings on the printer – for example, layer thickness and the build orientation, the mix of new-to-used material in a powder bed printer, and the storage of filament material – can all affect the outcome. These conditions and the 3D printer settings, therefore, need to be specified. Technicians frequently complain that build orientation is not specified when a part is to be 3D printed, and if the use of the product is not obvious, they are unable to judge which orientation the part needs to meet functional requirements. Where a part printed this way delaminates during use, for example, it is likely that the build orientation was not clarified. Who takes responsibility for this error, and for the costs incurred? Is it the designer or company who owns the product IP, and who sent the part to a printing bureau? Or is it the additive manufacturing company who printed the part that was sent to them? These issues are more easily controlled as part of a global networked company, but are increasingly difficult as the final manufacturers become separated from the company, such as a service bureau.

The unique characteristics of 3D printers as independent entities and the ability to send digital files between locations, combined with the rapid pace of development in digital communications, have also led to the emergence of 3D printing "build farms." Networked build farms enable a business model that supports a fully distributed system. 3D printing build farms are described as three or more similar printers networked to provide a larger scale service [7, 8]. One of the largest 3D printing build farms is run by Prusa Printers and has a Guinness World Record of 1096 printers running at the same time. The organisation produced masses of PPE and other supplies needed across Europe during the COVID-19 pandemic [9]. More sophisticated build farms incorporate autonomous post processing [10].

Where build farms could be a suitable solution to the challenges faced by businesses and communities across the globe, particularly in remote and regional centres [11], there needs to be a change in thinking and subsequent policy to support the practice. It does involve a very different way of looking at production and supply. Challenges such as the sharing of IP in a distributed manufacturing context, and accountability (which is the responsibility for the safety aspects in producing and using a product), will need to be more thoroughly and rigorously discussed and legislated for than they are now. Regulation informed by future-focussed research is needed, including on ways of tracking emissions.

For build farms to have an impact beyond emergency PPE production, they will need supportive infrastructure. For this, they will need to be seen as part of the smart city, as well as regional and remote living agendas. They have the potential to support decentralised living, allowing people to live and work in a digitally connected, but very socially distanced, environment. This could reduce the pressures on city living and support the changing values of the consumers as well as the needs of the prosumers, who are people who produce the products they need thanks to technologies like 3D printing, distinct from consumers who must purchase products [12]. For product designers, build farms and other distributed manufacturing models could help hasten the discipline's emergence from the constraints of conventional mass production into a new world of independent design and community production.

As discussed throughout this book, the confusion over the range of technologies under the additive manufacturing banner, and their distinct differences, causes problems when discussing any change to business practice. This is because what is appropriate for one technology is not appropriate for another. Distributed manufacturing, using additive manufacturing, can refer to the batch production of simple parts on relatively low-level printers, or the shared fabrication of a product over several printers, or complex products where intellectual property is shared between the networks. Increasingly, independent product designers are becoming involved in distributed manufacturing, producing work remotely, and either running the 3D printers themselves or working in collaboration with a fabrication network. Preparing the product design profession to work effectively in this way, and to show leadership in the development of changing business practices, is at a critical juncture.

There is an entrepreneurial streak within the product design profession that additive manufacturing enables in ways that were not possible before. The ability to work with distributed manufacturing networks, to utilise service bureaus to send digital files to customers overseas, and to create products not possible before because of manufacturing constraints is an opportunity for the profession. The adaptability of this way of working responds to some of the challenges facing the world currently. However, the discipline will need to support significant professional development to help product designers exploit emerging commercial opportunities through upskilling of the workforce. More broadly, a distributed manufacturing approach could form the backbone of a drive for greater equity in production practice and, as discussed in Chapter 8, support sustainability.

PERSONALISATION

Prosthetics for children tend to be expensive, partly because as the child grows, the prosthetics become ill-fitting and must be replaced. Whilst a complex functional prosthetic may be suitable for an adult, for a child the ability to protect the arm during play can be a priority. This was the case with the Child's Play prosthetic in Figure 5.5. Both versions of the prosthetic were modelled from a 3D scan of the child's arm so that the socket fitted perfectly. The 3D prints were designed to be lightweight, durable, and easy to clean, as well as replicable if ever damaged or lost.

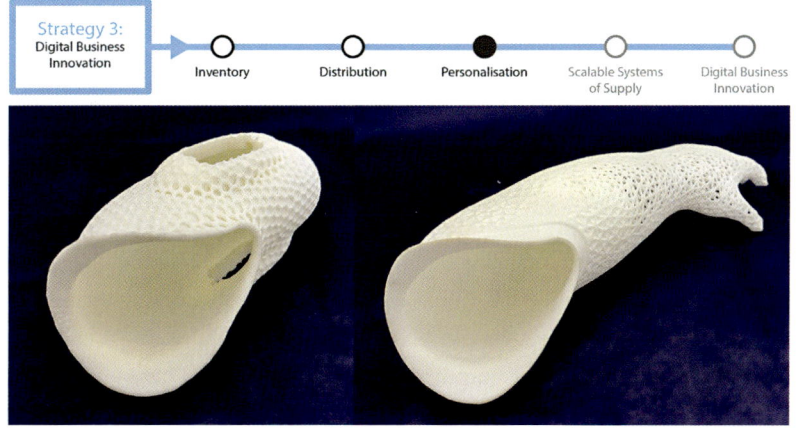

5.5
Example of personalisation – Child's Play prosthetic made to fit an arm scan (Diegel).

One of the easiest ways to understand the potential of this technology is to look at the case of Ziggy the dog, whose story made international news [13]. Ziggy was in a car accident as a puppy. He lost one leg, and the other leg was severely compromised. The surgery needed to correct Ziggy's remaining front leg was complicated. To assist the vet undertaking this procedure, CT scan data from Ziggy's leg was collected by scanning specialist and product designer Chris Little in the form of DICOM data (Digital Imaging and Communications in Medicine), who 3D modelled the bones in collaboration with the vet surgeon (a process known as "segmentation").

Product designers can play a role in this process because of their advanced CAD modelling expertise and ability to visualise in both 2D and 3D (noting that CT scans are actually many individual 2D slices through anatomy). The 3D model could then be 3D printed, allowing the vet to practice surgery in advance,

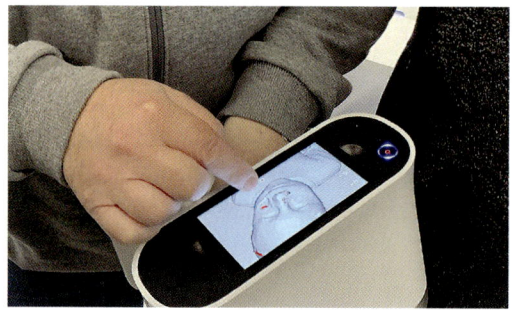

5.6
3D scanning interfaces are becoming easier to work with, and a wider range of equipment is now available.

checking on different approaches to fix the problem. Without 3D printing, this would be difficult to do, as producing one-off models without 3D printing technology would require considerable craft skills and would be expensive to repeat, especially if it was not suitable to produce using a mould. The outcomes for Ziggy were better than any expected, making a lifesaving change for Ziggy and Ziggy's owners.

Where 3D CAD was once used as a documentation and visualisation tool towards the end of the design process, developments in software, and associated technologies like 3D scanning, have shifted its role in the workflow for product designers to much earlier in the process. In fact, for many personalisation workflows, 3D CAD happens right at the start of a project, as a product designer converts scan data (slices or point clouds) into a 3D model as a base to begin working from. Various parametric or generative tools may then rapidly and automatically generate a design, without the traditional stages of concept development, and sketching product designers typically follow. The impacts of this on the forms favoured by the designers using such technology, and subsequent trends on product design, are yet to be fully explored.

This initiates not only new design workflows but, for businesses, also new interactions with users and collaborators who are stakeholders in the personalisation process. This can also include product customisation and manual interventions. However, unlike customisation, as defined within Strategy 2, in Strategy 3 this is supported by sophisticated business models necessary to support the "markets of one" described by personalisation. These markets are often niche and come with other unique challenges for a business. However, driven by the medical discipline, personalisation is becoming increasingly common as the benefits of 3D printed models for surgical planning, as well as other patient-specific devices like hearing aids and dental aligners, become clearer from both an economic and outcomes perspective.

For the product designer engaged by a company to extend a customisation strategy to the extremes of personalisation for additive manufacturing, the important shift to requiring investment and planning is that of going from constrained, often parametric, manual interventions that can be relatively easily predicted, to a much broader range of possibilities where the very first step inputs unknown geometry into the design equation. In many contexts, the system may require a full-time product designer or engineer who is involved in various capacities through the product personalisation process. This makes sense in many medical contexts where the cost of this can be offset by the rebates available in many countries for personalised medical devices, as well as the higher price of medical devices in general, compared to consumer goods.

Yet, the same business model may not apply to another industry, such as footwear. In this industry, personalisation strategies tend to be more automated and rely on combinations of technologies such as mobile phones with high-resolution cameras and Internet connectivity, cloud-based computing, machine learning, and build farms. The product designer sets up the system and validates an extreme range of use cases to ensure it works and can then move on to another project. The system lives on and "designs" products whenever a new input is provided. Computation replaces the costly time of a designer to personalise every product, which of course raises questions about how this will affect designers' perceptions of their roles, how they present their services, and how they are viewed by customers in the future. Instead of additive manufacturing being an additional technology to the current suite available to manufacturers, and therefore product designers, it can potentially create an existential crisis for the discipline, much as the rise of design thinking in business schools did. The extent to which the discipline is challenged by this will emerge over the next decade or so as software becomes increasingly capable of automating complex workflows that integrate personalised inputs.

DESIGNER'S COMMENT

With conventional manufacturing, a poorly designed and poorly manufactured part is about three times more expensive than a well-designed and efficiently manufactured part. With additive manufacturing, the difference can be tenfold – investing time and expertise in good design is even more critical. Understanding how to exploit the opportunities of additive manufacturing, whilst working within its constraints, is only possible with expertise both in the technical aspects of using the technology and also the challenges and potential it provides for business innovation.

SCALABLE SYSTEMS OF SUPPLY

Creating large frames for a variety of uses, for example from beams and connectors as in Figure 5.7, is nothing new. However, this system exploits the capabilities of 3D printing to provide a system where the angles and form of the connectors can be determined to a high degree of accuracy and produced on demand. This allows for unusual geometries to be systemised.

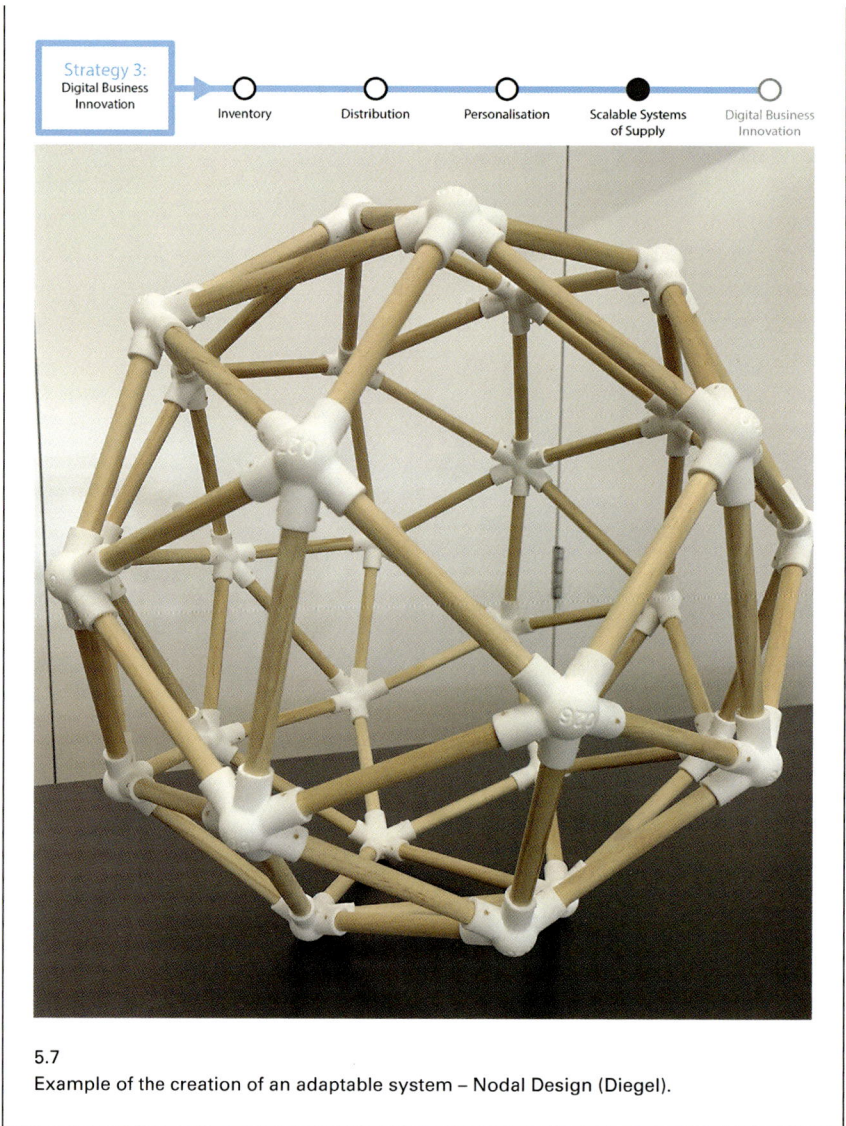

Strategy 3:
Digital Business
Innovation

Inventory Distribution Personalisation Scalable Systems Digital Business
of Supply Innovation

5.7
Example of the creation of an adaptable system – Nodal Design (Diegel).

In additive manufacturing, a doubling in size typically represents an eightfold increase in cost. This has largely restricted 3D printing to being used in smaller-sized parts. "Nodal design," however, is a highly appropriate design technique for cost reduction in large-scale AM projects and presents a range of new 3D printing possibilities. This can also be known as "tube and lug" or "hub and strut" construction. Nodal design involves transforming a 3D object into a set of nodes connected by rods to reconstitute the 3D object. Because the nodes are relatively small, they can be cost effective to print. And because every node is unique for organic shapes, 3D printing is the only sensible manufacturing method to make them. The more nodes are used in the model, the more accurate the overall shape becomes. A geodesic dome is a good example of nodal design, although in this case nodes are kept the same as much as possible to utilise traditional manufacturing techniques.

By its nature, nodal design creates parts that are structurally very rigid but lightweight. It can also be used as a technique for creating the wireframe "birdcage" that gets used to create the understructure for large-scale art projects. The "birdcage" can then be covered with materials such as canvas, plaster, and latex to create the final finished artwork.

The proposition of algorithmically engineered nodes and rods led the Creative Design and Additive Manufacturing Lab, based at the University of Auckland, New Zealand, to develop computational design workflows that automatically convert "solid" CAD designs or STL files into "node"-based designs. These comprise of 3D printed nodes, connected by rods, to form complex structures. The technique employs a relatively novel form of CAD modelling known as implicit modelling, a lightweight method of representing complex 3D objects using mathematical functions to describe solid bodies, making it highly adaptable to computational design, which is also formulae driven.

Of importance is the ability to sequentially label each unique node with its order and orientation, without which assembly would be too complex. Although originally intended for art projects, this computational design breakthrough has piqued the interests of researchers in architecture and construction. This highlights an interesting opportunity to overcome current limitations in structural engineering and construction. The project has, therefore, extended this computational design concept into the construction industry to efficiently build complex architectural structures that meet code and that can be quickly built without the need for specialised builders. This can be thought of as an "IKEA-style" construction method. The idea is to design the structure with nodes connected by I-beams or other connectors and then clip on panels for insulation, cladding, services, and so on.

The Lab has also extended this nodal design idea to other engineering applications, an example of which is the automated design of spaceframes, such as in Figure 5.8 for automotive applications. In this case, the University of Auckland Formula SAE race car team has used the workflow to create nodal-designed spaceframes for their race cars.

Often when a company decides to adopt additive manufacturing, they think about this in isolation for producing a new product or part. This can limit the scale of products to those suitable for common build volumes or require large gantry systems used for 3D printing concrete buildings. There is a middle ground where 3D printing is combined with other manufacturing methods that can accommodate customisation, allowing for larger and variable structures that may be at the scale of people or buildings. These will employ parametric,

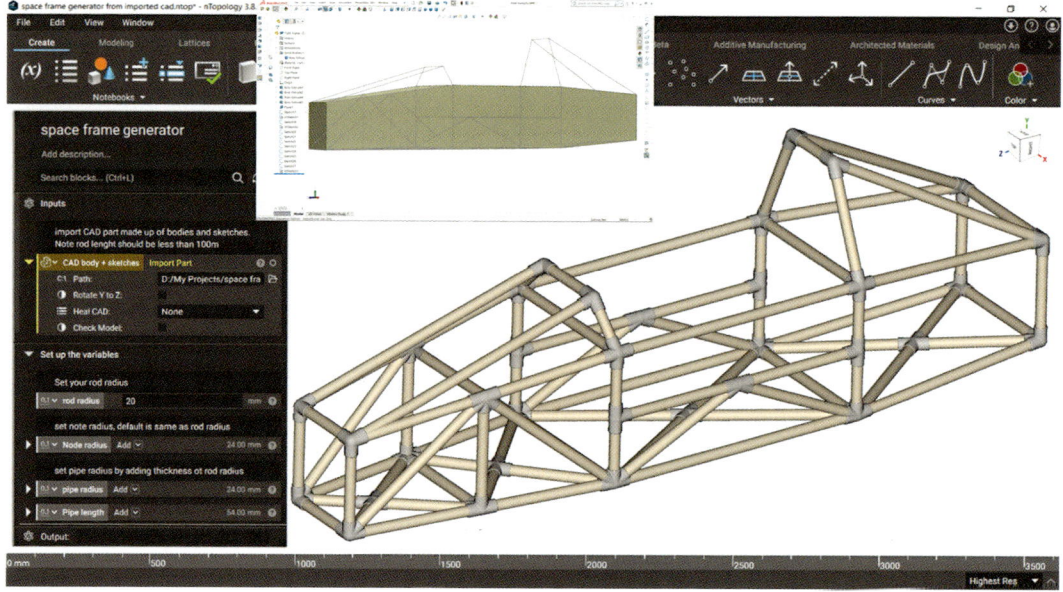

generative, or algorithmic software to generate new forms, as well as feed out the individual components in forms suitable for production. In other words, STL files of each custom connector are used for 3D printing, and CNC cutting toolpaths are used for plywood sheet components. Each part has an identifier to enable rapid assembly. These types of scalable systems build upon the digital distribution systems described previously.

For product designers, the increasing accessibility of high-end 3D printing for prototyping, and the speed it creates during the product development cycle, has been the key factor when working with clients. However, mass production itself is beset with inefficiencies. The ability to create bespoke end-use products from their inception through additive manufacturing, rather than through changes made further downstream in the process, requires a change in systems thinking. The COVID-19 pandemic arguably allows for the consideration of more radical innovation to business systems. Sales could be matched to production, with a reduction in lost sales because of an unanticipated surge in demand. Equally, overstock could be avoided.

Logistical postponement delays the decision to approve the production of a component in its final form for as long as possible. This happens in the fashion industry, where products are produced as "vanilla" – that is, colours are not added until the actual orders are received. The extreme version of this is termed "drop shipping," where products are not purchased by an intermediary or produced by a manufacturer until orders are received. This approach has increased with the rise in online shopping caused by the pandemic.

The contradiction between an increase in drop shipping, and the new ability to personalise products *ab initio*, through developments in digital technology, is

5.8
Example of an automatically generated race car spaceframe from a solid CAD model.

not yet resolved. Supply chain managers rarely have a good foundational knowledge of additive manufacturing. Working with product designers to respond to the potential impact digital files and supply can have on their operations is essential, just as IT specialists will need to work more closely with product designers to maximise the scalability of systems for customisation. There is little evidence, yet, of a good collaboration across these disciplines, but as product designers become more involved in the technology and its use in commercial practice, hopefully this will change.

DIGITAL BUSINESS INNOVATION

ProsFit is an EU-based company manufacturing prosthetic sockets and connectors and providing complete prostheses, as shown in Figure 5.9. Their products are compliant with EU directive 93/42EEC, approved by Certified Testing House Germany, and certified to ISO 10328. They are all currently up to a 125 kg load. The company uses 3D printing to reduce costs and improve the speed of production, the accuracy, and, therefore, the comfort of the sockets. They enable user-driven innovation in design and materials usage targeted at improving outcomes. What is particularly interesting about the company is the innovative systems approach they have brought into the field, aiming for scalability and global reach for their product and process. Their intent is to support the development of more universal access to their products and services, working with distributed healthcare networks to deliver digitally enabled medical services and high performing customised prosthetics more cost-effectively.

The key to their end-to-end system is that it is digital, with cloud-based software platforms that enables scan data to be uploaded. This then drives the customised design output, which is then bespoke manufactured using 3D printing. Although ProsFit is agnostic to which 3D printing technology is used, today it uses mainly HP's Multi Jet Fusion (MJF). ProsFit has built up a global network of manufacturers with HP's machines, allowing the parts to be produced as near to the customer as possible. ProsFit has also developed a design and expert system to be used by clinical professionals that deploys CAD design, digital biomechanics, finite element analysis, machine learning, and digital solution architectures. With this approach, ProsFit is demonstrating how to exploit emerging technology, and make a digital manufacturing solution scalable in a digital health care context. This is a very good example of digital business innovation, responding to the opportunities of digital convergence. The work fits into Strategy 3, described in this book, and illustrates how 3D printing is not a stand-alone technology but part of a digital suite that enables new ways of working.

5.9
Example of digital business innovation – Prosfit, led by Alan and Christopher Hutchison.
(Images courtesy of Prosfit.)

In the current era, companies need to be agile, as their average life span is rapidly decreasing. They need to be aware of threats and opportunities provided both by the emergence of new technologies and by trends in society. Issues such as climate change, pollution, the rise of automation, and Industry 4.0 can all impact a business, particularly as governing bodies and political policies respond. To stay ahead, they need not only to work in their business on a day-to-day basis, but also to ensure their businesses are responsive to the changes emerging from its business environment. The example of Kodak's bankruptcy so soon after making such large profits is a case in point, exacerbated by their failure to see the opportunities of digital photography despite owning several relevant patents. In contrast, leading tyre manufacturer Michelin continues to look to the future, as shown in their futuristic scenario where 3D printed airless tyres from recycled materials are "regrown" as needed. In the meantime, they have recently developed an airless tyre that, whilst not as futuristic as the 3D printed ones, does demonstrate innovation in practice.

For effective additive manufacturing integration, the strategies discussed through this chapter and the preceding Strategy chapters allow for a methodical approach to working more effectively in each new industry sector. The final approach – that is, the culmination of those outlined through Strategy 3 – is digital business innovation. The definition of digital business innovation is not about providing tools for working on the web or with social media; it is about value adding in response to the opportunities disruptive digital technologies provide. This chapter discusses, for the product designer, the unanticipated consequences of disruptive change, and prepares them to work with a company on making significant changes to maximise the opportunities provided by the technology. Building upon Strategy 2, it works through issues around the relationship between the customer and the producer, and its potential for change, and the operational impacts of introducing 3D printing into an existing business. This chapter is also important for entrepreneurs and start-ups whose agility allows them to begin working at the cutting edge of additive manufacturing unencumbered by traditional manufacturing paradigms.

Additive manufacturing business innovation examples

For product designers, rapid prototyping as a professional service was available long before desktop printers were made readily accessible around 2010. However, 3D printing bureaus were themselves expensive, and the service they provided was frequently linked to design and development. In 2007, a new departure emerged, with a company in the Netherlands called Shapeways that came out of Royal Philips Electronics. This company provided an online, user-friendly, 3D printing service and a retail platform for individuals to share their work. The bureau grew rapidly, investing in production-scale machines that could be optimised to produce maximum output. The online interface was developed to be accessible to the amateur as well as to the professional designer. This opened new markets, including an online shop for designers to sell their products via the Shapeways website without having to hold any inventory or own any manufacturing equipment themselves. Print-on-demand could service both manufacturers and customers.

The hearing aid company Phonak altered its business practice to a personalisation approach, changing its interactions with its customers by scanning the inner ear of each customer to produce bespoke outcomes. These hearing aids are printed in batches, but each one is individual to the customer. A similar model has since been adopted for dental aligners [14], which are typically vacuum formed over personalised 3D printed models from 3D scans of patients in retail-facing stores. Software automatically plans for the staged movement of teeth based on the original 3D scan, and the patient is sent a kit of clear aligners that progressively moves their teeth over time. What was once the domain of highly trained orthodontists, who assembled metal braces and adjusted them at regular check-ups, is quickly becoming automated through 3D printing.

In conventional manufacturing, there are considerable hidden costs and delays when the whole supply chain and production system are considered [15]. The argument is, therefore, that any comparison between additive manufacturing and conventional manufacturing must take these inefficiencies into consideration. GE Additive provides a useful example to illustrate the extent of business change possible through a shift to additive manufacturing. GE Additive is one of the world leaders in additive manufacturing, and their work over the last ten years has led aerospace in the appropriate adoption of additive manufacturing. However, more significant for a systems approach to the evaluation of additive manufacturing as a commercially viable proposition, their work clearly demonstrates the reason for studying this technology as opposed to other manufacturing technologies at this time. The impact of design for additive manufacturing on the whole business structure and functions in GE Additive has been dramatic. Additive manufacturing has the potential to bring added value and efficiencies to most companies, if used appropriately and designed as a system rather than as a stand-alone product.

A recent project from GE Additive is their advanced turboprop engine. In this engine, hundreds of parts from their traditionally manufactured and assembled engine were consolidated into tens. The new additively manufactured engine resulted in a significant weight reduction and fuel burn reduction. It also resulted in a reduction of the critical combustor test schedule. These are clear and what appear to be quite straightforward positives. However, what the team highlighted, themselves, was the unexpected impact that they had not anticipated on their operations. This related to the drastic reduction in the number of design files and their management, and the number of manufacturing sources. Inspection systems could be rationalised through the approach, repair sources could be reduced, and the data systems involved were also dramatically reduced.

This administrative reduction was unexpected and could potentially cause some serious change management issues for any company fully adopting an additive manufacturing strategy like this. Long term, it would allow a company to redeploy personnel, save money, and increase output. But it does illustrate that additive manufacturing, done thoroughly, can create a very different business model. This is significant when thinking about business practice and the implications, good or bad, of introducing additive manufacturing.

Customer interaction

When considering maximising the opportunities that additive manufacturing provides, and adding value rather than replacing an existing product, the ability to create personalised, or at least customised, products is high on the list of priorities. Comparing a like-for-like product made using conventional manufacturing with one made using additive manufacturing rarely shows additive manufacturing as a viable alternative. Yet, where additive manufacturing clearly adds value is in a product design service system, where customisation (Strategy 2) or personalisation (Strategy 3) is involved. However, a key factor that needs to be understood and prepared for by a company working on this approach is the necessity to integrate new ways of interacting with the customer to make this a reality. If, for example, an automotive company decides to offer a customisation element to the finishes in their cars, as BMW did, then it needs to build a platform to enable that to happen. This is not only a cost for the company that needs to be considered at the start; it also will require ongoing services and expertise. Anticipating the problems this may cause needs to be part of adopting a customisation or personalisation strategy.

There are numerous approaches to building a customer interaction platform. This may involve a platform based on a digital library of parts for the customer to choose from, or it may be more innovative, such as based on an animation that the customer can stop at a point in time to select the outcomes. The animation is less controlled, with examples such as Fung Kwok Pan's Fluid Vase [16] having a disclaimer attached that the selection may not print properly. With parametric examples such as those by Nervous System, the range of motion achieved in the 3D model is constrained and verified so that any selection within it can be verified. However, as innovative as such examples may be, for many businesses today it is important to highlight that these are not new cutting-edge examples. The technologies, and research, to create online interactive systems are well established and can be applied to almost any product type.

For the product designer working with a company that has existing skills in using additive manufacturing, and that is comfortable with the process they have selected, moving on to proposing an online platform or other method of creating a product service system over a product, with customer interaction inherent to the system, takes the company to the next level of digital business innovation with additive manufacturing. Understanding issues such as the shift from working with consumers, to consumers and prosumers, the use of connectivity and pre-emptive practice in creating product service systems, closed loop production, and the future of industrial practice is still ongoing. Additive manufacturing can also allow the design of radically different value-added innovative products that are likely to involve a change in business model.

DESIGNER'S COMMENT

What is the value of time? It is almost universally acknowledged that AM can reduce lead times and change the relationship between producer and consumer, helping to bring more appropriate products to a particular market faster. Does the increase in profit, by coming to market early, justify the increase in part cost or tool cost from AM, or the cost of retraining staff and changing the customer interface?

For the future of the discipline, there needs to be a shift in emphasis from product output to workforce development. Product designers need to be supported in developing the skills to work effectively with the technology. There is a danger, otherwise, that a new offshoot of information technology might emerge to undermine the discipline. The whole tenet of demand and supply is changing, and the role of product designers will change with it as digital technology continues to impact the economy. Anticipating the next decade and preparing for what might happen to the discipline is a challenge affected by influencers and policy makers.

WHAT MIGHT HAVE HAPPENED AT THE MULTINATIONAL COMPANY …

After researching the market opportunities with LND Consulting, the board of the fictional multinational company Hydnora Technologies elected to make a significant investment in working with digital technologies and split their core business into two. The new branch of the company, ironically led by the initially reluctant Clare Torres, created a global distributed manufacturing network to contract the supply chain of high-end lightweight parts for small aircraft and automotive applications. The business model focussed on critical parts and need-it-now situations. They also found new markets in the mining and public transport industries because of their ability to supply rapidly. In the mines, critical parts were needed ASAP to keep operations running, and they required flying in, which had an impact on the cost of transportation. The major rail networks in several countries also began using the company to provide critical parts to keep their operations running, often for trains and systems that were several decades old and difficult to find parts for.

The change in operations for the company meant new ways of working, new training for the workforce, and very different supply chain distribution and monitoring systems. As part of development, they launched a product service system approach where breakages could be anticipated, and parts supplied automatically. The profile of the company improved, and elements of the approach began to be adopted in the original branch of the company.

DESIGNER'S COMMENT

Thinking beyond a single digital technology to working with a suite of technologies and integrating ideas on connected products and Industry 4.0 is a challenge. Using techniques such as backcasting helps a company to imagine where it would like to be in five years (for example), then work backwards. This can be a good starting point for digital business innovation, rather than thinking about existing practice as the basis for innovation.

THE END OF THE JOURNEY … OR IS IT?

The technology adoption journey described in these three Strategy chapters provides product designers and companies with a model of practice that works through additive manufacturing technology adoption step-by-step. By the end of the journey, a company should have built a body of knowledge and new operational systems to maximise the opportunities the technology provides. However, in terms of design for now, working to address the growing sustainability imperative is arguably the next stage of the journey. Chapter 8, on 3D printing sustainability, provides starting points for this next stage of technology adoption.

REFERENCES

1. Harford T. *Messy: How to Be Creative and Resilient in a Tidy-Minded World.* London: Little Brown; 2016.
2. Loy J, Tatham P, Healey R, Tapper CL. 3D Printing Meets Humanitarian Design Research: Creative Technologies in Remote Regions. In: Connor AM, Marks S, editors. *Creative Technologies for Multidisciplinary Applications.* Hershey, PA: IGI Global; 2016. pp. 54–75.
3. Mohr S, Khan O. 3D Printing and Its Disruptive Impacts on Supply Chains of the Future. *Technology Innovation Management Review* 2015;5(11):20–25.
4. Novak JI, Loy J. A Critical Review of Initial 3D Printed Products Responding to COVID-19 Health and Supply Chain Challenges. *Emerald Open Research* 2020;2(24).
5. Tino R, Moore R, Antoline S, Ravi P, Wake N, Ionita CN, et al. COVID-19 and the Role of 3D Printing in Medicine. *3D Printing in Medicine* 2020;6(1):11.
6. Novak JI, Loy J. A Quantitative Analysis of 3D Printed Face Shields and Masks during COVID-19. *Emerald Open Research* 2020;2(42).
7. Laplume A, Anzalone GC, Pearce JM. Open-Source, Self-Replicating 3-D Printer Factory for Small-Business Manufacturing. *The International Journal of Advanced Manufacturing Technology* 2016;85(1):633–642.
8. Loy J, Novak JI. 3D Printing Build Farms: The Rise of a Distributed Manufacturing Workforce. In: Blount Y, Gloet M, editors. *Anywhere Working and the Future of Work.* Hershey, PA: IGI Global; 2021. pp. 220–246.
9. Research P. Prusa Face Shield: Prusa Printers; 2020 [updated 26 March 2020. Available from: https://www.prusaprinters.org/prints/25857-prusa-face-shield.
10. Ransikarbum K, Ha S, Ma J, Kim N. Multi-Objective Optimization Analysis for Part-to-Printer Assignment in a Network of 3D Fused Deposition Modeling. *Journal of Manufacturing Systems* 2017;43:35–46.
11. Krywko J. 3D Printing Farms: A Glimpse into the Future. *Zortrax* 2018 [updated 20/9/2018. Available from: https://zortrax.com/blog/3d-printing-farms-a-glimpse-into-the-future/.
12. Ahluwalia P, Miller T. The Prosumer. *Social Identities* 2014;20(4–5):259–261.
13. Lieu J. Three-Legged Ziggy the Dog Is Walking Again Thanks to 3D Printing. *Mashable* 2016. Available from https://mashable.com/article/three-legged-dog-3d.

14. Davies S. Smilelove & SmileDirectClub Discuss the Benefit of 3D Printing Moulds for Clear Aligners. *TCT Magazine* 2019. Available from: https://www.tctmagazine.com/3d-printing-news/aligner-start-ups-show-their-teeth-smilelove-smiledirectclub/.

15. Lipson H, Kurman M. *Fabricated: The New World of 3D Printing*. Somerset, NJ: Wiley; 2013.

16. Archer N. Fung Kwok Pan: Fluid Vase: Designboom; 2010. Available from https://www.designboom.com/technology/fung-kwok-pan-fluid-vase/.

6 Case studies

3D printing from the product designers' perspective

CASE STUDY 1: TOUGHER AND FASTER: ADDITIVE MANUFACTURING FOR BICYCLES

Sports products are a very good starting point for the use of additive manufacturing for product designers for several reasons. First, the money available in the industry to explore the potential to go faster, be safer, and create personalised products across many sports is vast [1]. There are good examples of practice in the use of additive manufacturing in creating more advanced mouthguards, and additive manufacturing is being used in research into protective headgear because of the ability to print lattice structures that can absorb impacts, fitting around 3D scans of an athlete's head. This case study section focusses on bicycles, as there are examples of practice across all three stages of technology adoption, and this industry has been at the forefront of product design using the technology for the last decade and continues to innovate.

Bicycles are an iconic product that have evolved over two centuries, adopting and informing new manufacturing methods and materials in the pursuit of improved performance, safety, comfort, and aesthetics. While often seen as a single product, a bicycle is an assembly of many different sub-assemblies and products which operate as one, including the bicycle frame, pedals, saddle, handlebars, gears, wheels, and brakes. Alongside these are numerous supporting products to aid in maintenance, transportation, comfort, smart connectivity, and safety, making for a large ecosystem of cycling products. The industries' embrace of new manufacturing technologies has seen 3D printing rapidly gain traction throughout the cycling ecosystem. While it has been used for several decades as a prototyping tool, it has expanded to include many of the other strategies outlined in the previous chapters. As a result, it is a good starting point for considering where product designers are advancing the use of the technology in industry.

Prototyping and bridge manufacturing of bicycle accessories

The cycling product market is highly competitive and dominated by large multi-national brands. For small companies and start-ups, it can be challenging to break into this market and make the large capital investments needed for traditional manufacturing. Rapid prototyping and bridge manufacturing with 3D printing can help decrease the time to market and lower the barriers to entry. This results in batches of end-use parts for the market to test before committing to mass production, as shown in Figure 6.1. Rehook is an award-winning product that is used to fix a dropped bicycle chain back onto the sprocket, saving the rider from getting oily hands. The product began as a side project for Rehook founder Wayne Taylor, who was frustrated after dropping his chain on the way to a meeting and turning up late, covered in oil. He used a desktop FFF 3D printer to prototype designs and test them on his bike. This allowed for rapid iteration and refinement of the idea.

After finalising a design, Rehook needed a batch of 50 products that had mechanical and aesthetic properties closer to a final manufactured part for further testing. They chose SLS, utilising a service bureau to fabricate the parts. Testing of this batch revealed a design weakness that was not observed with FFF and allowed the Rehook team to refine the design; if this had already been an injection-moulded product, such a discovery would have been a very costly mistake to correct. Material experiments for SLS 3D printing were also done, with initial carbon-reinforced nylon swapped to graphite-reinforced nylon to reduce the weight of the product. This was an important development to maintain part strength while considering the needs of cyclists, with this tool either carried on the bike or by the cyclist.

6.1
A selection of Rehook prototypes 3D printed using desktop FFF to allow quick and cost-effective iteration (a); SLS production of Rehook in progress (b). (Image Credit: Rehook (a); Alexei Burton of 3Dprintdirect.co.uk (b)).

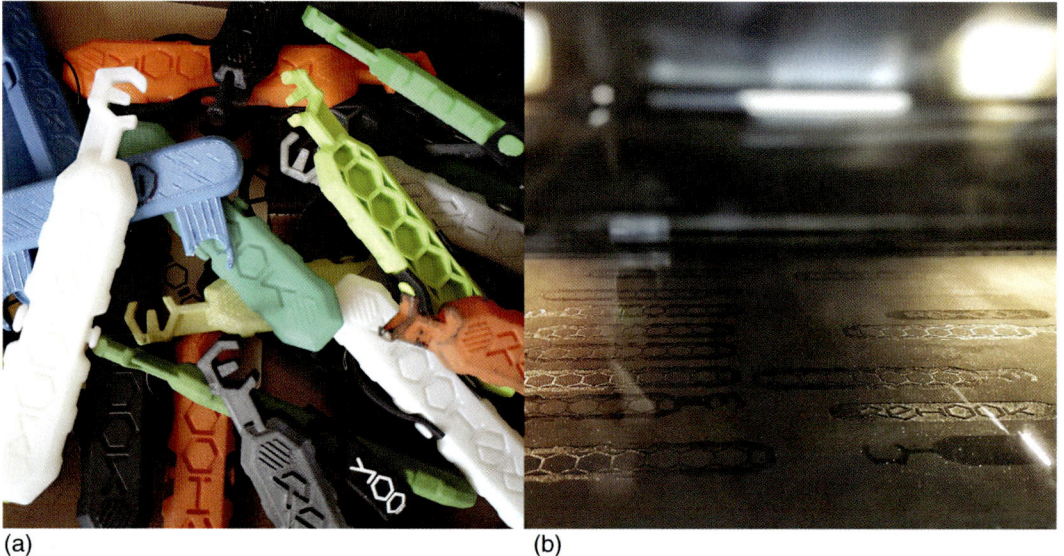

(a)   (b)

6.2
Evolution of Rehook from prototype to final product. FFF prototypes (blue and green), through to SLS batch production (middle grey and black), and final injection-moulded design (right). (Image Credit: Rehook).

While some companies would shift to mass production at this point, Rehook decided to continue using SLS 3D printing for batch production, testing the market before committing to any further upfront manufacturing costs. Their SLS supplier produced 400 units per month at an affordable rate by fitting them into their existing production runs. This allowed Rehook to bring products to market quickly and begin their marketing campaign, illustrated in Figure 6.2. The total timeframe from initial idea through to taking a first batch of end-use products to a cycling trade show was just ten weeks. This was for a total cost of US$1200/1000 pounds sterling, with only 50 percent of this cost spent on developing the tool. The other 50 percent was initial marketing. Rehook sold out at the first trade show, and that gave the company the confidence to scale up production. This meant that they shifted to injection moulding within 12 months of the initial idea. This is a pathway that many small businesses, start-ups, and even large-scale manufacturers can adopt to embrace the rapid pace of design innovation balanced with the desire to sell large volumes of product in a market as large as the cycling industry.

Optimisation of lightweight bicycle components

Manufacturers of bicycle frames and components are often driven to adopt new materials or manufacturing methods to deliver ever more lightweight products. Reduced weight can mean it is easier to accelerate off the starting line, peddle uphill, or manoeuvre over rough terrain. For this reason, the ability for 3D printing to produce complex geometries that traditional subtractive or moulding technologies could not produce provides new opportunities to reduce bicycle weight.

Bicycle manufacturer Empire Cycles, located in Bolton, UK, collaborated with the manufacturer of selective laser melting (SLM) metal 3D printers, Renishaw, located in Wotton-under-Edge, UK. This basis of the collaboration was to design, and 3D print, a full mountain bike frame. One of the key elements of the design was the new seat post bracket of the frame [2]. This was a useful starting point to explore the capabilities of additive manufacturing due to the small size of this part compared to the full frame, which was suitable to the limited build volume of SLM machines from Renishaw at the time. The design of the seat post bracket was informed by topology optimisation software that identified regions of material that were not necessary to meet desired mechanical performance. Critical mounting holes and the tube for the saddle were restricted from changes.

This resulted in an organic model between these critical regions, with features that could not normally be manufactured through traditional methods. In this case, the optimised model was used as a template by designers at Empire Cycles, who created clean geometry following the principles learned from topology optimisation. This step also took into consideration some of the constraints of SLM, modifying some angles so that the seat post bracket would print with minimal support structures.

The seat post bracket shown in Figure 6.3, as well as the full bicycle frame sectioned into smaller pieces, was 3D printed in a titanium alloy using Renishaw's SLM technology. For comparison, the original aluminium alloy seat post bracket from Empire Cycles weighed 360 g, while the optimised titanium bracket weighed 200 g, which is a weight saving of 44 percent. Collectively, the full 3D printed frame weighed 1400 g compared to a traditional frame of 2100 g, a 33 percent weight saving. The seat post bracket was tested to the mountain bike

6.3
Process of topology optimisation of the seat post bracket with a final image of the bracket in its printed orientation, amongst other frame pieces on a build plate. (Image Credit: Renishaw).

1. CAD model of seat post designed for aluminium alloy casting

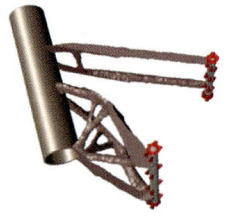

2. Topological optimisation using Altair's solidThinking Inspire® 9.5 software

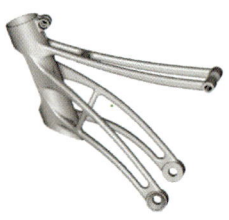

3. Redesigned by Empire Cycles using the optimised CAD model as a template

4. Produce in titanium alloy on a Renishaw AM250 laser melting system

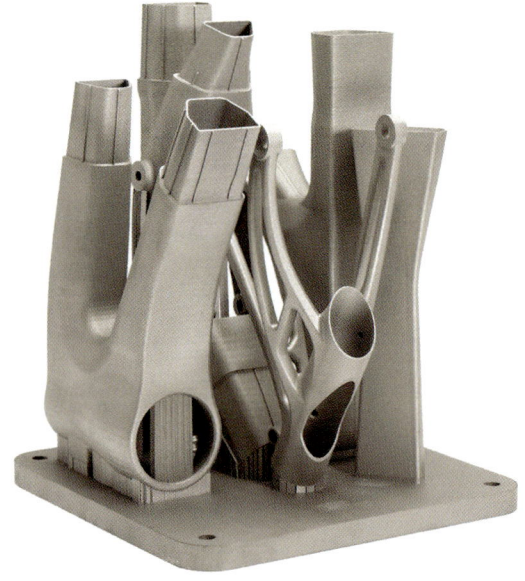

6.4
Complete 3D
printed titanium
bicycle frame from
Empire Cycles and
Renishaw. (Image
Credit: Renishaw).

standard EN 14766, shown in Figure 6.4, and withstood 50,000 cycles of 1200N, continuing to six times the standard without failure.

> **DESIGNER'S COMMENT**
>
> Metal additive manufacturing is the most expensive form of manufacturing in the industry. It should be used only when it adds enough value to overcome the high costs.

This project demonstrates how advanced CAD software enables designers to embrace simulation processes within the design process, rather than as a separate process that occurs after design. The geometry freedoms afforded by 3D printing allow designers to utilise processes like topology optimisation to create forms that would previously not be manufacturable. It is important for this end manufacturing process to be determined prior to design to take advantage of these opportunities. Many developments are happening with 3D printed bicycle frames and components, which is likely to further drive innovation in this industry.

The future of 3D printed bicycles
The reality for 3D printing is that materials and processes suitable for the requirements of a bicycle frame are limited in scale, meaning that direct 3D printing a complete bicycle frame as one part remains challenging today. However, this does not mean it is impossible, and just as concept cars help shape the future of the automotive industry, concept bicycles are important for shaping the future of the cycling industry.

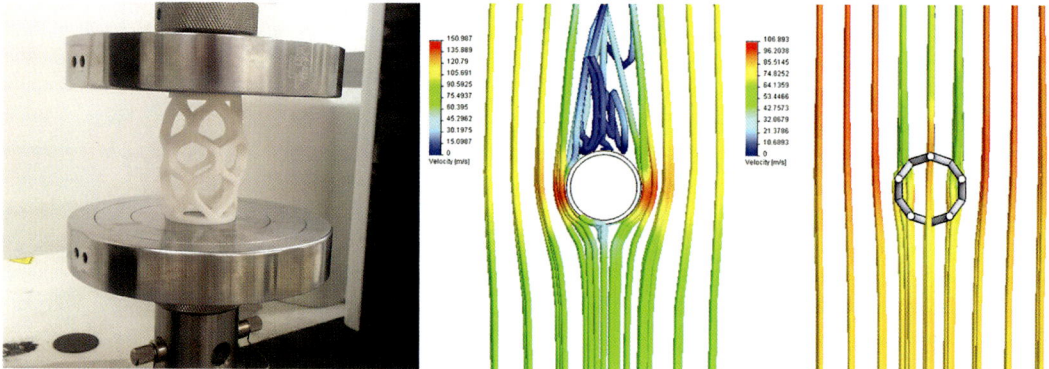

6.5
Compression
testing of early
lattice sample (left),
comparison of
aerodynamic
properties of a
standard tube and
lattice geometry
using simulation
(right).

The FIX3D bicycle frame by Dr James Novak was one of the first to reimagine what a fully 3D printed frame could look like. The project began with months of experiments and tests using a range of 3D printers, including both desktop and large commercial FFF machines, and material jetting technology [3]. This allowed for rapid iteration and learning about how to design for additive manufacturing using lightweight lattice structures. Test pieces were put through compression testing and aerodynamics analysis procedures, shown in Figure 6.5.

The final FIX3D bicycle frame was created as a parametric system in Solidworks, taking 150 hours to model. The parametric nature of the model meant that cyclist measurements (anthropometry) could be input into the model, and the bicycle frame would rebuild itself using these dimensions. This included the height and placement of the saddle, consolidating several separate parts that are normally required for bicycles to suit anyone who buys them. As a result, this also reduced bicycle weight, as did the use of a lattice structure for the overall design.

Other features of the project included a personalised badge forming part of the structure of the bicycle frame, in this case, using the designer's last name. Safety features were also considered, with rear-facing LED lights accommodated within the frame structure. This made safety a built-in feature rather than an optional extra. The design, FIX3D, shown in Figure 6.6. has been exhibited around the world at venues like the Red Dot Design Museum in Essen, Germany, and the BOZAR Centre for Fine Arts in Brussels, Belgium. For several of these exhibitions, Dr Novak has simply emailed the STL file of the frame to the organiser, who could 3D print the frame locally then assemble the frame out of accessible bike components rather than require the complete bicycle to be shipped between venues. This may be the future for cycling events!

Cycling (and other) helmets

Since 2017 there has been a significant growth in interest in the use of 3D printing for protective gear in sports. In terms of 3D printing innovation, one of the key developments has been the work of Carbon and Riddell on the

6.6
FIX3D bicycle frame (Novak).

Speedflex Diamond Helmet (https://www.carbon3d.com/riddell) for American football. The reason this is a key development is because it exploits the capabilities of 3D printing to build lattice structures and rethink the form of the helmet itself [4]. The Riddell helmet is precision-fit; that is, the geometry is determined by a scan of the athlete. The design of the helmet is data driven, being informed by data about on-field impacts, and the generative design software runs through over a thousand options to create the optimal design for that player. The print is created using Carbon's Digital Light Synthesis (DLS) technology to build an end-use product from resin. The nature of the technology allows the density and geometries within the lattice structure to be altered as needed. Therefore, it demonstrates the future for product designers. They will be able to realistically manipulate the very fabric of a product to meet the performance, or aesthetic, requirements they need. In addition, it highlights the growing integration of IT with product design over conventional engineering. This is a challenge for higher education, where new pathways and collaborations need to be built, and rapidly.

3D printing is being used, or under development, by a range of companies for other protective gear in sports, such as face masks and mouth guards for different contact sports. Hexr is a company operating in this space, producing bicycle helmets based on 3D scan data and 3D printed to fit. The company is aiming more at the mass market with scan data collected on iPhones by the customer. Even so, the company states that the helmet is built on a 250,000-point 3D model of the customer's head to ensure an accurate fit. The company also claim that the impact absorbency of 3D printing allows for a 26 percent safer product than one made using conventional foam.

However, 3D printing also allows product designers to rethink the use case for products, rather than mimic those already in existence. This is not only through the ability to create bespoke products, but also sophisticated prototypes that demonstrate the bringing together of multiple advanced technologies. This is demonstrated in a concept bicycle helmet that was a finalist in the wearable technology design competition, Reshape 19. "Dynaero" challenged the ruling of Union Cycliste Internationale (UCI) that helmets should not be equipped with any technology features. Designed by Dr James Novak, the design argues for a more user-centred approach to ensure that cyclist's health is a priority, and that the potential to improve the aerodynamics of the wearer through intelligent design should be viewed, much as in Formula 1, as an extension of the competition.

During endurance races such as the Tour de France, cyclists must choose between ventilation and aerodynamics. Dynaero is electromechanically able to respond to temperature, speed, and other real-time conditions to open and close vents in the helmet independent of the rider. During wind tunnel tests with a cycling mannequin (shown in Figure 6.7), aerodynamic performance was improved by 3.7 percent between the open and closed positions of the helmet [5]. There are also considerable prospective health benefits of being able to

6.7
Wind tunnel testing
of Dynaero at 44
km/h (Image credit:
Timothy Crouch).

6.8
Dynaero 3D printed
bike helmet and
accompanying
mobile app (Novak).

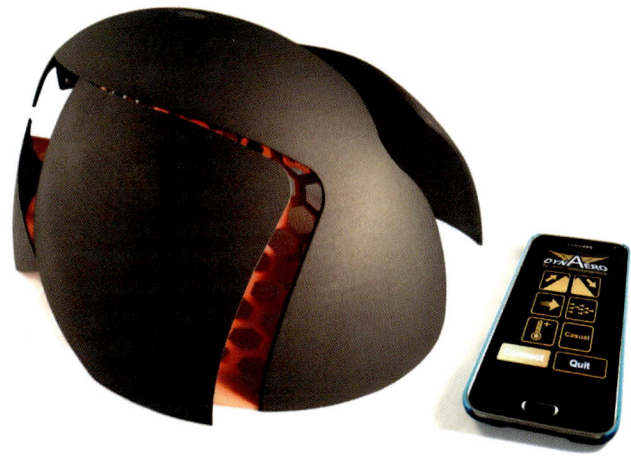

automatically control the ventilation during intense riding conditions. The proto-type, shown in Figure 6.8, was 3D printed using powder bed laser sintered nylon, and the final product would be custom-fit based on 3D scan data.

With the Dynaero helmet, the mobile phone app was connected via Bluetooth to the prototype to act as a sensor for controlling helmet ventilation, meaning that a cyclist did not need to wear any other sensors or connect other devices to the system for early tests. The project demonstrates the shift in thinking described through Strategy 3 of this book, where 3D printing can be a catalyst to challenge the status quo. Similar advances may be seen across many other sports as the digital increasingly invades the physical world, and athletes, coaches, sponsors, and even recreational players look for the competitive edge.

CASE STUDY 2: ADDITIVE MANUFACTURING OF MUSICAL INSTRUMENTS: GUITARS

Throughout the history of musical instruments, manufacturing techniques and materials have continuously evolved. Clarinets, for example, have been made from wood, metal, Bakelite, and other plastics, and some saxophones are now made from injection-moulded plastic, rather than the more common brass. Some brass instruments are also now being manufactured in plastics. The evolution of manufacturing techniques and materials has been driven by several factors, including acoustic qualities, aesthetics, cost, availability of materials, and ease of manufacture.

For ODD Guitars, a New Zealand company that manufactures 3D printed electric guitars, it began as pure experimentation and exploring the possibilities of 3D printing in musical instruments [6]. After their very first solid body electric guitar was printed, the added value of 3D printing quickly became apparent. Chief amongst these advantages were:

- The ability to create a unique guitar aesthetic, with incredibly complex geometries, that would not be possible to produce with conventional manufacturing methods.
- The ability to easily customise each instrument to suit the style and aesthetic preferences of each individual musician, which added great value to the instruments. For ODD Guitars, no two guitars are ever the same.
- As the designs are digital, it is easy to use software to see the centre of gravity of the guitar and then digitally move material around to give it a heavier or lighter neck feel depending on the player's preferences. This also allows the guitars to be made a little bit lighter than conventional guitars. There was a limit to this, however, because if the body were made too light, the guitar would become "neck-heavy."

This added value was incredibly important as, to be commercially viable, the value that was being added by using 3D printing, which is a relatively expensive manufacturing method, needed to be high enough to overcome those high costs. After producing their first electric guitar in 2011, ODD Guitars did a blog post on the instrument, and the response from musicians around the world was instant; orders for instruments started coming in. To date, ODD Guitars has manufactured ninety-three 3D printed guitars, so around ten guitars a year.

The Atom, depicted in Figure 6.9, was one of the first commercially produced electric guitars from ODD Guitars. It was printed in nylon using powder bed fusion (SLS) and had the stair-stepping layer lines left purposefully visible on the curved top surface, similar to a wood grain effect. Later, Atom guitars were made with a smooth front surface. In the context of using additive manufacturing in the construction of musical instruments, two main factors are critical to the quality of the instruments produced: the mechanical characteristics of the instrument, and the acoustic characteristics. Whatever material is used, it must

6.9
The Atom (Diegel).

also be safe to play from a toxicological point of view (although this is less critical for guitars compared to other wind instruments, such as those involving contact with a player's mouth).

Mechanical requirements: The prime consideration of an additively manufactured instrument is that it be strong enough to withstand the rigors of playing without getting deformed or damaged. On stringed instruments such as guitars, for example, there can be over 100 kg of tension on the strings, which means that whatever material and structure is used to manufacture the instrument, it must be sufficient to withstand that amount of tension without deforming to the point where the instrument is no longer playable, or its intonation is no longer correct. In this respect, the stronger, and more rigid, the 3D printing technology and material being used, the better. Some currently manufactured 3D printing instruments also use a hybrid approach in which they combine additive manufacturing with conventional materials such as wood. Stringed instruments have been successfully manufactured using vat photopolymerisation, powder bed fusion (both polymer and metal), and material extrusion-based, additive manufacturing technologies.

With additively manufactured guitars, for example, ODD Guitars undertook trials with different materials and geometries to find the most suitable combination for an electric guitar. In this case, the instruments were manufactured using powder bed fusion out of polyamide (PA2200/nylon). In the initial design configuration, the guitar body was made of solid nylon and had a small "waist" between the pickups.

Finite element analysis in Figure 6.10 showed that there would be a deflection in the original body design of up to 2 mm with the strings under 100 kg of

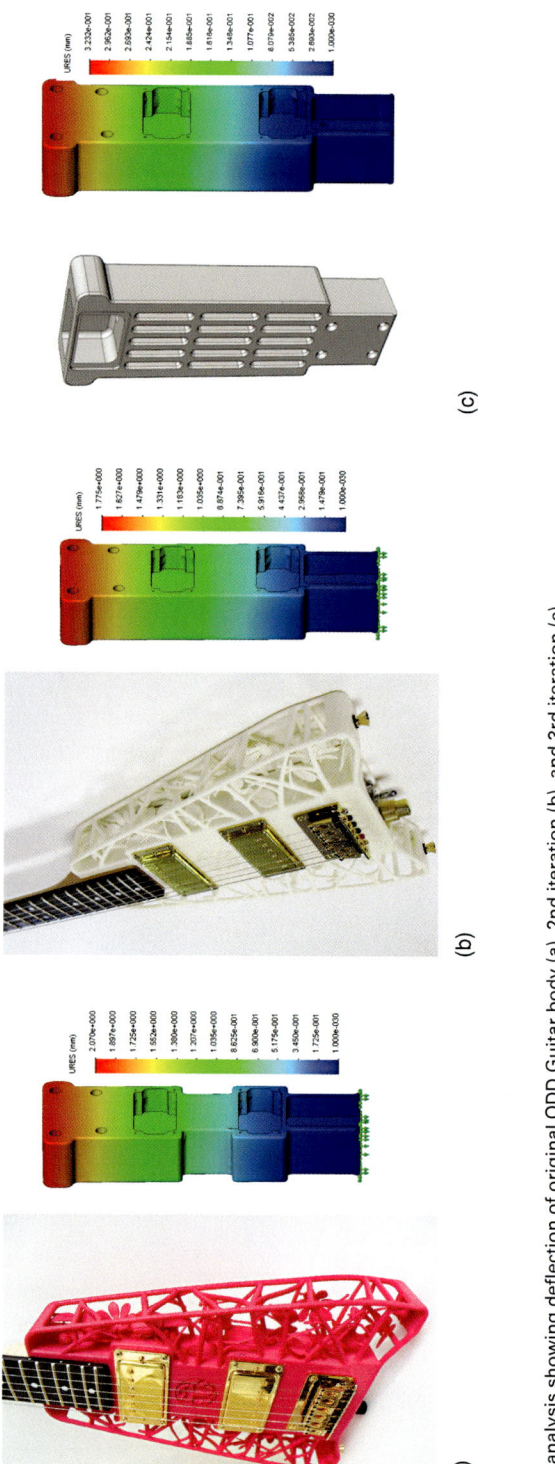

6.10

FEA analysis showing deflection of original ODD Guitar body (a), 2nd iteration (b), and 3rd iteration (c).

tension. A second design iteration was tested with the "waist" removed, and this reduced the deflection to 1.7 mm. A third iteration was tested in which the body was filled with strengthening ribs to increase the rigidity of the body. In this design, the deflection was reduced to less than 0.3 mm, which was deemed acceptable for an electric guitar. However, after consulting with the market and having the instruments tested by several musicians, ODD Guitars found a perception problem with a solid nylon body, in that it was subjectively perceived as inferior to a wooden body. For this reason, the guitars were redesigned to contain a wooden core inside the additively manufactured nylon body. Effectively, this meant that, if the nylon body was entirely removed, the guitar would be a standard wooden-bodied electric guitar, albeit with a small wooden body. This solution served to address both any mechanical issues with a 3D printed guitar and, at the same time, also addressed some of the perceived acoustic issues which will be discussed in the following section.

Acoustic requirements: There is a never-ending debate between guitar aficionados about the benefits, or not, of tone-woods. For an acoustic guitar, in which the body is the main resonance chamber that produces the instrument's sound, the material, undoubtably, plays a major role in the instrument. But, for a solid-body electric guitar, the material used for the body has only a very minor role in its sound. The pickups, or a fresh set of guitar strings, for example, has an infinitely larger influence on the sound of the guitar than the body material does. A musician named Jim Lill has produced an excellent video that clearly demonstrates where the sound of an electric guitar comes from. This can be found through this link: https://www.youtube.com/watch?v=n02tImce3AE.

Nevertheless, from a marketing point of view, just as in the endless debate between the qualities of music listened to on vinyl versus digital, it would have been very hard to convince the tone-wood believers that printing a guitar in any alternative material would meet their standards. This was the other reason why ODD Guitars put a wooden core inside the 3D printed body, joining the neck and the bridge to, effectively, create a small-bodied wooden guitar with a 3D printed skin. This meant that, from a marketing point of view, the guitar could be advertised as having a "soul of wood." And for increased customisation, the musician could choose what flavour of tone-wood (maple, mahogany, etc.) was used for their guitar core.

A brief history of 3D printed electric guitars

Several electric guitars have been produced using additive manufacturing technologies. The earliest of these appears to have been produced with an SLA system. This was designed by Owain Pedgley and Eddie Norman at Loughborough University in 2005 and is currently owned by Dr Ian Campbell. There have also been several guitars produced by Tim Thellin, of RedEye/Stratasys using FFF technologies. FFF has also been used by Derek Manson of One.61 in New Zealand to produce an electric guitar. An experimental music guitar/bass was also created by Ziv Bar Ilan, called the Zoybar, which is commercially available and uses SLS as its manufacturing technology. The Zoybar is available through the online service bureau Shapeways. In 2010, an international contest was held in collaboration with Shapeways for anyone to build on the Zoybar original CAD files and customise the design however they liked.

The first commercially available electric guitars and basses were those produced by ODD Guitars in 2011. They were manufactured with powder bed fusion (SLS) in New Zealand and featured a nylon body, with an inner wooden core joining the bridge to the neck. The approach used allowed for customisation of hardware and shape by the customer (as in Figure 6.11). ODD Guitars has also produced two electric guitars with a 3D printed aluminium body. It is important to note that, on all the commercially available 3D printed guitars to date, only the bodies have been 3D printed; the necks are made from wood, and the rest of the hardware are manufactured using conventional manufacturing technologies.

6.11
A production run of Americana and Atom guitars. Each guitar has features specific to each musician (Diegel).

SLS nylon produces extremely strong parts, but they require post-processing and painting to give them whatever aesthetic properties are requested by the customer. The painting process can be extremely time-consuming and therefore expensive. An example of this is the Beatlemania bass guitar. The guitar is themed after Paul McCartney's iconic Hofner violin-shaped bass guitar, and the front panel is made up with the score for "Yesterday." The inside contains various Beatles-era-related icons (Yellow Submarine, Sgt Pepper's Lonely Hearts Club Band drum kit, George Harrison's Rickenbacker guitar, John Lennon's glasses, the iconic Abbey Road scene, etc.). As the guitar was printed in white nylon, painting all the internal and external details was a painstaking manual process that took around three weeks. Because of the considerable effort required to paint the instruments, ODD Guitars began to experiment with printing with some of the available full-colour material jetting 3D printing technologies. The first results produced with full-colour printing not only were visually stunning but also reduced the labour time to produce a guitar by several weeks. The only post-processing that was needed for these guitar bodies was to dissolve the soluble wax support material, and then clearcoat the instrument body with a UV-resistant clear acrylic paint to protect it from ambient UV light.

The body of the Scarab ST guitar in Figure 6.12 is not painted after printing. All the colour, including the wood grain, is printed by the material jetting system. It was printed on a Mimaki 3DUJ-553 in acrylate resin. ODD Guitars has also manufactured two guitars that were 3D printed in aluminium using metal powder bed fusion technologies, shown in Figure 6.13. As metal 3D printing requires support material to prevent the parts from distorting, these guitars required a substantial amount of post-processing labour to remove this support material, and sanding, filing, and media-blasting to get the surface finish to an acceptable level.

6.12
The Scarab ST full-colour 3D printed guitar (Diegel).

(a)

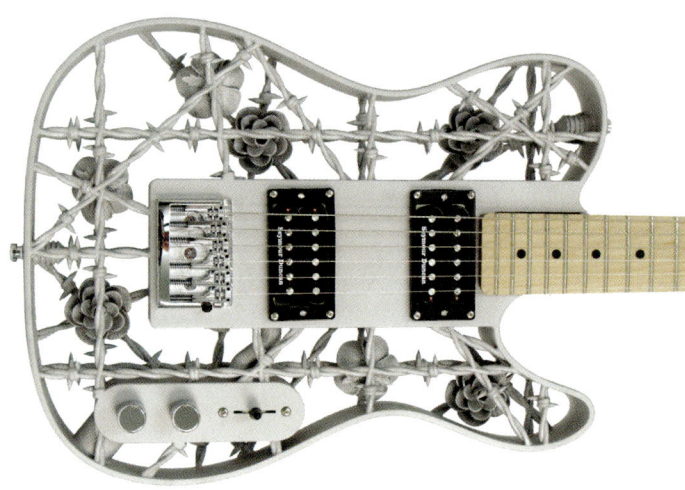

(b)

6.13
Aluminium guitar still welded to the build-plate, and close-up showing the support material below the barbed wires (a). Finished Heavy Metal aluminium 3D printed guitar (b) (Diegel).

The most recent ODD Guitars design involved experimentation experimenting with new materials and testing of a new binder-jetting technology, made by a US company called Forust, that prints in wood. The process used sawdust and inkjet prints lignin-based bio-epoxy onto it to join it together. In other woods, the process uses a source of waste material (sawdust and lignin) that would otherwise be incinerated or go to landfill and transforms them into high-value products through a binder-jetting 3D print process. The process can used different colours of bio-epoxy to mimic the wood grain patterns of wood, shown in Figure 6.14.

The guitars described to this point are relatively high-end custom guitars that are made using high-end additive manufacturing technologies. It is for this reason that they benefit from the customisation and aesthetic complexity allowed by 3D printing. Over the past few years, there have also been several examples of fully 3D printed guitars, from body and neck, produced using low-cost material extrusion (FFF). However, these guitars are often printed without the required hardware (truss rods, etc.) needed to make a guitar both adjustable and durable. So, without extensive further testing, it is hard to judge whether those guitars will be of a quality to compete with traditionally made guitars. However, as lower-cost, entry-level hobby guitars, there may well be a market for such instruments, especially if a musician can produce their own guitar and value the freedom to do this, over the need for higher-end materials/finishes and acoustic properties.

6.14
The GreenAxe 3D printed wood guitar. Printed in sawdust and lignin-based bio epoxy using a proprietary binder-jetting 3D printing process (Diegel).

Industrial designer Adrian McCormack demonstrated his ability to create bespoke guitars in his commissioned work for the 2016 Blues on Broadbeach Festival in Australia, undertaken during his final year of study. The overall aesthetic of the design was influenced by the Gold Coast's rich surfing culture and history, whilst maintaining a body shape typical of a Fender Telecaster guitar and aiming for a similar weight to that of the Telecaster. However, it was very much a bespoke item, incorporating design details to respond to both the client and his own aesthetics. The design features breaking waves throughout the sections of the body, with a rising sun silhouette in the background of the design providing a nice contrast to the waves and several supporting "beams" which added strength to the overall shape. Other smaller "easter eggs" are hidden within the design, including references to the late Gold Coast surfing icon Michael "MP" Peterson and the name and date of the festival.

McCormack used design detailing for additive manufacturing in the work, creating variations of his guitar for different additive manufacturing processes. The white guitar, shown in Figure 6.15, was 3D printed in a single piece in a large format selective laser sintering (SLS) machine. The project had a short timeline, six weeks from concept sketching to finished product. This included a minimum of two weeks for the selective laser sintering of the guitar by I. Materialise in Belgium, and its delivery to Australia.

DESIGNER'S COMMENT

With 3D printing, it doesn't cost any more to make your product beautiful and add value through building in practical features to your design. In fact, cutting logos, part numbers, instructions, etc., into your part reduces the amount of material used and therefore the cost.

6.15
Close-up of Adrian McCormack's SLS guitar.

6.16
Marbled finish achieved using hydro-dipping for the SLS printed guitar (McCormack).

The one-part SLS print needed specific design considerations implemented into the 3D model to optimise the design for SLS 3D printing. The guitar was clear coated to avoid any discolouration of the polyamide material. A later version of the SLS printed guitar was painted using a "hydro-dipping" technique, shown in Figure 6.16, to achieve a very organic and totally one-of-a-kind finish.

The build volume was a constraint in both cases, but particularly the red-and-white version which was an altered design to suit a dual-filament FFF machine with a build volume of just 250 mm cubed. The guitar was instead sectioned into seven pieces with multiple dovetail joins used in the design to connect the parts together. These needed to be hidden from the front view of the guitar and can be seen only from the back view. A tolerance of 0.2 mm was given between these joins to allow for any possible defects from the prints, as

6.17
Multiple piece
guitar, filament
printed
(McCormack).

well as space for adhesive to hold the parts together. The total print time of the FFF parts was around eight days, printing in 0.1 mm layers with a nozzle size of 0.17 mm, meaning extremely high-definition prints but a longer print time than "normal" FFF printing. The supports were PVA material and dissolved in a turbulent heated bath of caustic soda.

The three guitars by McCormack were fitted with necks and hardware by Rohan Staples at The Guitar Shop in Paddington, Brisbane. Staples said it was the first time he had worked on guitar bodies which had been 3D printed, made from plastic polymers, and was interested that it arrived ready for all the components to be attached as opposed to having to position and pre-drill the body for attaching components to it. The final weight of the SLS printed body in Figure 6.15 was 1.6 kg, and the FFF printed body guitar in Figure 6.17 was 1.2 kg. This weight difference is largely due to the use of infill structures with FFF printing, introducing air cavities, whereas the SLS design was essentially solid.

As in many industries in which aesthetic product design considerations and customisation play a big role, musical instruments are a good example of an industry in which 3D printing can be used to add considerable value to the instruments being produced. It is also an example of where product designers can create their own product lines, using 3D printing service bureaus, to make innovative products for individual clients.

CASE STUDY 3: 3D PRINTING FOR THE DESIGNER-MAKER: FURNITURE

Designing furniture is a classic product design – and architectural – obsession. Most students will have designed an item of furniture during their studies, and many design professionals will have at least one piece of furniture on their resume. Designer-makers are even more likely to make furniture, and to explore the relationship between the handmade and machine-made furniture using emerging technology. In the first 20 years of this century, the one area where product designers led in design for 3D printing has been in furniture. Results tend to be experimental rather than commercial, but they do demonstrate very different ways of thinking about the use of the technology.

The designer-maker and the relationship between industrial product design and craftsmanship is a constant feature of the product design discourse. From the work of Pye on the role of the handmade [7] in design to the challenging ideas led by Lionel Dean, whose process is always evolving [8], product artistry informs the development of the discipline. The exploration by Ross Lovegrove, Ron Arad, and many other leaders in the industry that pushes the boundaries of what is possible enabled by 3D printing has challenged the discipline from both practical and theoretical points of view. In all cases, the personal imagination and connection with the creation of products is a constant.

Product designer and University of Technology Sydney academic Berto Pandolfo chose to shift the focus on design for 3D printing from the machine aesthetic to its relationship to the handmade. Having an industry-research background, Pandolfo initially worked on design for 3D printing for manufacturing. However, in 2017, he explored the tension between emerging and traditional manufacturing methods in his "MND" furniture series. The work combines handmade elements with 3D printing and questions the need for post-processing

6.18
Berto Pandolfo designed the "MND" furniture series exploring the relationship between the handmade and machine-made, in timber with 3D printed legs (image courtesy of Berto Pandolfo).

when using filament extrusion 3D printing and the assumption that 3D printed parts should be geometrically "perfect."

Whilst the timber cabinets in Figure 6.18 are finished to a high level, the 3D printed legs retain imperfections. The legs are built from rounded stone shapes, stacked to provide asymmetrical supports to contrast with the clean lines of the woodwork. The uneven, handcrafted aesthetic of the stacked stone legs is contrasted by their repeatable, highly mechanised system of production through fused filament fabrication. The pebble supports of the structures can be made in a range of sizes, and the ridges on the curves of the shapes are not polished out, filled, or smoothed, but stay in place to communicate how they were made to the user, much like the grain of the timber. The work illustrates the unique position of product designers, integrating craft and manufacturing, the personal and mechanisation.

Personalisation

Product designer Dr James Novak was similarly interested in bringing the personal into a technical process. Although his work was arguably the theme of "handmade," it had a very different starting point. Novak's 2019 Fingerprint Stool shown in Figure 6.19 explored the personalisation capability of additive manufacturing [9]. The geometry on one side of the design was derived from the fingerprint on James's ring finger, while the other side was the fingerprint of his fiancée's (now wife's) ring finger. The project began as an exploration of how to design and optimise a piece of furniture for large-format FFF production. Designing for this technology included specifying a flat base, as this would allow it to sit on the build platform without the need for support material. The goal was to ensure the stool could be produced in less than five days so that it could be monitored by lab staff during a standard working week and be produced for a budgeted price of US$1000/870 pounds sterling.

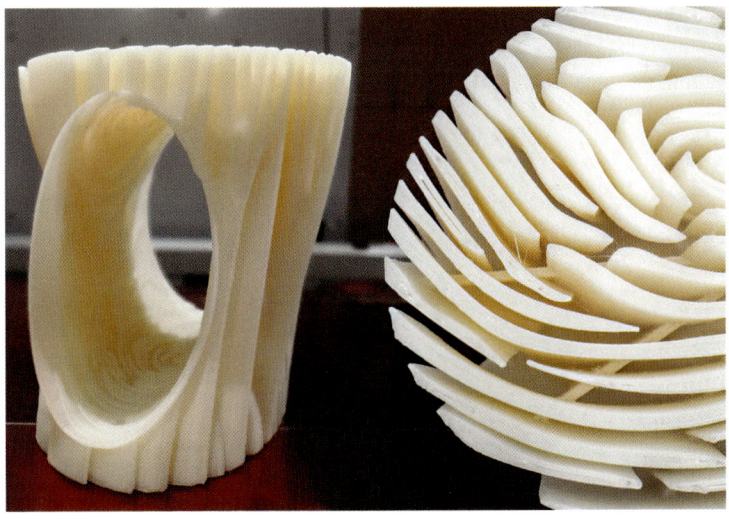

6.19
Fingerprint Stool
(Novak).

The design process involved taking impressions of the fingerprints with ink on paper, and then digitising them on a flatbed scanner. These were then vectorised using Adobe Illustrator and exported as DXF files. These were imported into Solidworks 3D CAD modelling software and oriented with a 420 mm gap between them, which defined the height of the stool. The geometry between the two was manually modelled, using lofts and sweeps where appropriate. An elliptical extrusion was used to cut through the centre, removing the mass of material (reduced weight and cost) and exposing the internal intricacies of the model. An ellipse was chosen since it can be 3D printed with FFF technology without support material, whereas a circle is more challenging and would require supports at this scale. The STL was then exported and sliced in Simplify 3D for production on a BigRep ONE, which has a build volume of 1005 × 1005 × 1005 mm.

Slicing revealed that the original design was too big and did not meet the criteria of print time or cost, no matter what process parameters were used. Novak had to iteratively trim the design into a narrower piece to achieve this. Prototypes were tested at small scales using desktop FFF machines to help. However, the first full-scale prints of this design were unsuccessful, as shown in Figure 6.20. Bed adhesion was challenging, and as the nozzle moved between the individual sections of the fingerprint, they were easily dislodged if there was any warpage or plastic sticking up. At one point this led to a nozzle being destroyed as the PLA plastic got caught up and encased the nozzle. Eventually, a raft was used to temporarily merge all the separate pieces of the fingerprint design and increase bed adhesion, which later took over one hour to remove using a hammer and chisel. This was the only waste material produced from the final print.

Considerable experimentation was also needed to refine the final process parameters, with Novak finding that at this scale, simply changing the number of perimeter walls from two to one could reduce the printing time by several

6.20
Failed print after being dislodged from the build plate without raft (a) and completed stool in the BigRep ONE printer (b).

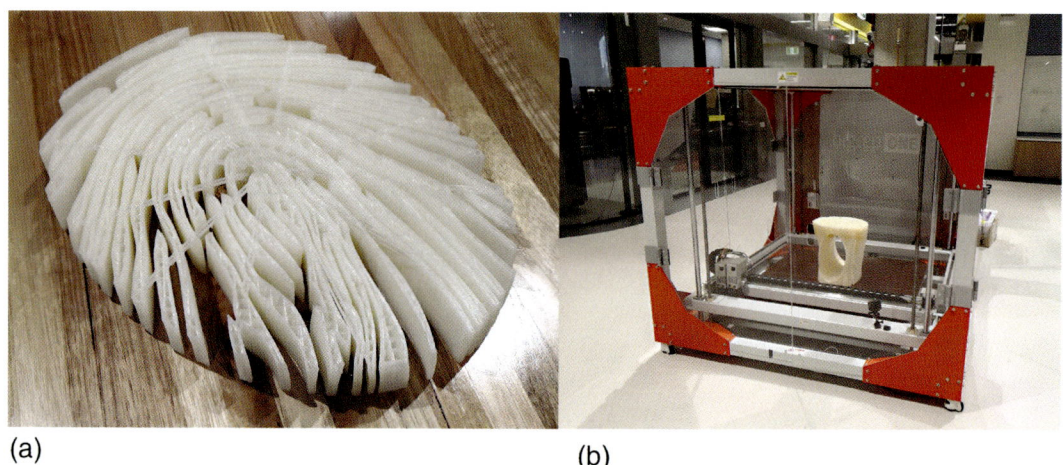

(a)   (b)

days. Eventually, a 1 mm diameter nozzle and a 0.5 mm layer height with 5 percent infill was specified to 3D print the stool, balancing strength requirements with printing time and cost limitations. The stool took 114 hours (3.75 days) to print and nearly 11 kg of PLA. The print layers are clearly visible in the final stool, but, much as with Pandolfo's MND work, Novak's view was that this added to the character of the stool and speaks to the method of production. Therefore, he chose not to look at any smoothing options.

New ways of working – furniture and IT

Whilst there are multiple good examples of design innovation for one-off or short-run pieces using additive manufacturing, the product design discipline is yet to develop a clear point of view on systems design for 3D printing. In general, the furniture being produced by practitioners in the discipline tends to be experimental in terms of the opportunities for innovation provided by the processes and sits within the parameters of Strategies 1 and 2 in this book. However, the work of leading product designer Lionel Dean at FutureFactories.com exploits the technology from a very different perspective and suggests a way forward into Strategy 3, considering new ways of developing systems for additive manufacturing.

Dean's work is unusual because its foundations from 2002 are built on developments in gaming and animation [10]. He describes himself as a product artist working at the creative boundaries of the product discipline and works exclusively with 3D printing. Dean's work can also be controversial. His choice of iconography in his Faberge limited editions, for example, subverts cultural norms and values. He challenges established practices in the discipline and was one of the first designers to build fluid design outcomes. He describes his work as producing "living products," rather than static outcomes, as his use of 3D printing for fabrication allows him to produce a physical outcome from the multiple versions of an idea that he generates through his process. His presentations argue for a very different way of creating products, and a different way of relating to those products that is at odds with established practice. Dean's work is considered culturally significant and has been acquired by numerous museums and art galleries, including the Museum of Modern Art in New York and the Design Museum Barcelona for their permanent collections.

Dean's work is significant for product designers not only because it challenges makers and consumers to rethink their relationship with products, but also because it illustrates a point that is frequently missed in the use of additive manufacturing. That is, although geometrical complexity can be 3D printed, designing a product on that basis alone does not exploit the most critical capability enabled by additive manufacturing, which is the ability to provide unique parts for every print. The underlying structures that enable the product designer to work with this facility need to be in-built before they can be used on a case-by-case basis. Dean illustrates this approach, using animations and storytelling as the basis for his work. By exploring the creative boundaries, Dean demonstrates the need to engage with products in new ways to fully exploit the technology and

6.21
Ghost Chair (Lionel
T. Dean,
FutureFactories).

for the designer to act as a "metadesigner" [11], who creates the environment for determining the design, rather than designing a resolved, single product or even range of designs.

> The emphasis on "counter-narrative" is important as it suggests that it is some-how different from the main narrative, either that which is explicitly and collectively agreed upon by society as being "mainstream" or being implicit in accepted behaviour (the underlying paradigm). The implication is that design activism voices other possibilities than those that already exist with a view to eliciting societal change and transformation.
>
> [12]

The Holy Ghost Chair, depicted in Figure 6.21, was a significant project by Dean. This was created in 2006, at a time when very few product designers were exploring additive manufacturing in depth. In this project, Dean specified the boundaries of a generative design space and the growth pattern needed for its design features. This was a computer-script-driven meta-design, generating numerous design iterations for the same theme, but with different detailing. The boundaries provided were the size of the space that the algorithm could fill with nodes, and different growth patterns that could be used to achieve this. The final designs were 3D printed in polymer, then electroplated.

Overall, embracing the step-change innovation this approach enables, ensures new technologies are not overlooked in terms of affecting its development. As a relatively young discipline, it can create a paradigm shift in practice to respond to the changing values and ideas in the world that would be more difficult for more established disciplines, such as engineering. Now is the time for product designers to invest in the future of the discipline for a digital era.

CASE STUDY 4: NEW WAYS OF WORKING: FILM INDUSTRY – ANIMATION, PROPS, AND COSTUMES

Using the three strategies and their subcategories as outlined in this book as a starting point, it is possible to evaluate an industry sector, such as the special effects industry, to identify opportunities to develop new outcomes when using additive manufacturing. This will support the development of new ideas, rather than solely the reworking of existing products to be made by additive manufacturing.

Examples of where 3D printing is used in the film industry:

Rapid prototyping – making visual models for planning purposes.

Mass customisation – creation of multiple parts with slight variations; for example, for costumes or stop-motion animation, using parametric models.

Bridge manufacturing – making replica products for destruction, such as a section of a valuable car. This can also help with issues such as transport miles, where a product may be housed overseas and expensive or fragile to move. A scanned model, 3D printed and finished in the correct location, will avoid transportation – and insurance – issues

Light-weighting – props that need to be easily manoeuvrable, such as swords for fight scenes, which may look larger than life but are very light to handle.

Product innovation – Innovation forms that would be difficult to produce otherwise, such as the 3D printed headdress created for the film *Black Panther*.

For some time, leading forms of material jetting have enabled the creation of 3D printed colour objects that would be extremely difficult to produce using conventional techniques. Material jetting is based on large numbers of print heads, usually around 800, that "print" the part in tiny droplets, much as a standard 2D printer used in an office would, but with a gel-like material. This process allows colour to be used throughout the print. This is a perfect technology for the film industry, as the parts produced tend to be temporary. In other words, the material may not be suitable for long-term, end-use applications, and the colour

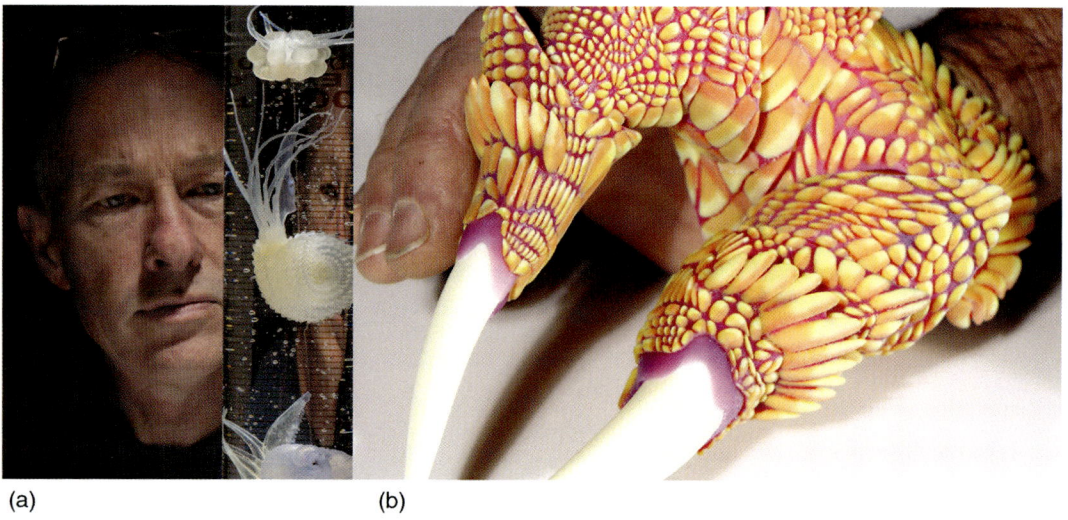

(a)   (b)

also tends to fade during exposure to light. However, for short-term use, such as for props, this technology is extremely useful. More recently, HP launched multi jet fusion, which also uses large banks of print heads in a 2D print-like process, though in this case the droplets enable or inhibit powder to absorb heat. This process also allows for colour to be added to parts, though in this case the colour is applied to the surface of the part rather than running through it, and the company was still exploring the viability of the colour version of the technology at the time of this writing. The interior of the part is a grey colour.

6.22
Pneumatic, flexible water-creatures by Ross Stevens (a) and rigid 3D printed prop glove by Tor Robinson and Ross Stevens of Victoria University of Wellington, New Zealand (b) (photos courtesy of Ross Stevens).

The advantage of this process is that the parts produced are much stronger. Post-processing can create a more resistant surface, and it is likely that the film industry will embrace this technology for applications over the next few years. However, one key advantage of the Stratasys polyjet process is that it is possible to adjust the transparency and flexibility of an object, as shown in Figure 6.22a. This is a challenge for the HP fusion process, although flexible material, such as TPU, can be used. The ability to create flexible, translucent parts is critical to visual production work for film, especially for 4D printing. For a range of 3D printing technologies, from FFF, SLS, and MJF to DLP, flexible materials can be used, such as TPU. In the innovative example by Ross Stevens, shown in Figure 6.22a, pneumatics is used to articulate the water-creatures, suitable for filming.

The 3D printed eyes shown in Figure 6.23 illustrate the potential of the technology to create convincing props for the film industry. In this example, the eyes were modelled without scanning. However, a scan is possible, both of details such as the eye, and for full bodies, and could be used to add accuracy in representing actors as physical models. This accuracy in representation was explored further by Olaf Diegel in a project to replicate his own face, shown in Figure 6.24.

6.23
Material jetting, 3D
printed, multi-
material eyes
(Diegel).

6.24
3D printed faces
based on 3D scan,
with 3D printed eye
inserts – see Figure
6.23 (Diegel).

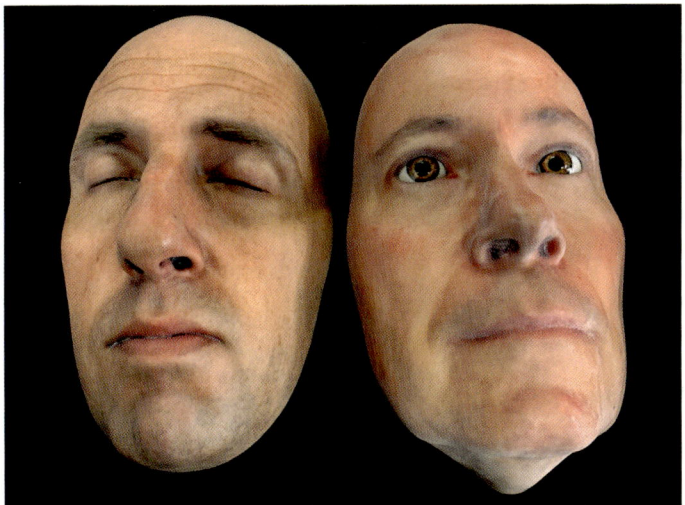

As the technologies evolve, new companies have begun to create colour material jetting processes that are potentially more robust and less expensive to run. Whilst these are still relatively untested, the early results are good, with machines capable of 3D printing more usable products suitable for the film industry that are starting to be of interest to large companies, such as Weta Workshops.

Changing practice

The film industry is an interesting example to consider in relation to the adoption of 3D printing, partly because it operates to the side of mainstream product design and is not constrained by the same need for mass production as a commercial practice. Also, it is interesting because it has so dramatically changed practice within companies that adopted 3D printing, and over a relatively short time. Understanding an industry sector well enough to see the opportunities with the technology is needed, but frequently those who could identify the opportunities are reluctant to engage with the technology as, on the surface, it replaces their existing practice and expertise. When introducing 3D printing into a workflow, it is not about replacing like-for-like, but in identifying where it can enable different ways of working and extends practice rather than replaces it. The film industry provides an interesting example of a sector that has developed unevenly in terms of the adoption of 3D printing, with a selection of lead-user companies illustrating the opportunities it can provide.

When the US special effects company Legacy Effects bought their first 3D printer, they had in mind that it would be a useful tool for the early visualisation of props and costumes, such as through maquettes. However, once they bought it, they found it was useful for making moulds and scale models for props and costumes too. Soon they began making parts for costumes out of 3D printing. The accessibility and bespoke capabilities of the technology meant they were able to produce individual parts for a specific need within the film, in a way that was not previously viable.

3D printing changed not only how they could make props and parts for costumes, but also how a scene could be filmed. For example, the first Iron Man suit for Robert Downey Jr was expensive, and there was a limit to what could be filmed whilst he was wearing it because it could get damaged in a fight scene, or it was heavy, or it offered limited movement. Even where moulds could be made from the original, parts were delicate and easily broken. This restricted what they could be used for on the film set. However, they began to make modular parts for the suit using 3D printing, which meant they could put the suit under greater duress, as parts could be replaced as needed.

In the later Iron Man films, Legacy Effects could build multiple suits using 3D printing, and those suits could have built-in parts that were customised to the use they were required for. For example, where a suit was being used in a stunt, a part may be designed to withstand additional impact, or to compress or shatter, if that was more appropriate for the stunt. Since Legacy Effects started to use additive manufacturing for prototypes over a decade ago, it has increased the range of additive manufacturing technologies it employs as the understanding of the production engineers working at Legacy Effects increases. From visual scale models to secure approval of a product concept, through 3D printing moulds, to end-use 3D printing for props, mechanisms, and costumes, Legacy Effects can now not only replicate parts that it created previously through other means, but also can build new product outcomes. These can impact how the company operates in unanticipated ways.

Another example of changing practice and the extension of practice enabled by 3D printing in the film industry, rather than just replacing practice, comes from stop-motion animation. In the 1993 stop-motion animation film *The Nightmare before Christmas*, the main character, Jack Skeleton, was made with

700 hand-sculpted heads to allow for the range of expressions during the film. If a 3D printer was brought into the workflow, it might be expected that a single printer would be able to produce those 700 heads fairly rapidly, replacing the work of a team of sculptors. The parts would still need to be digitally sculpted, still requiring craft skills, albeit digital ones. However, when Laika Stop Motion Animation Studios in the United States brought in 3D printers in 2009 for the movie *Coraline*, they were able to make 20,000 unique faces for their main character instead of 700. In 2012, they made 33,000 heads for the film *ParaNorman*. By 2014, in their film *The Boxtrolls*, they were making 50,000 heads to provide the animators with a full range of expressions for the stop-motion characters. Instead of replicating existing practice, 3D printing technology allowed for an expansion of practice.

Boxtrolls heads were made using non-colour binder jetting technology. For this technology the outcomes would be white, and still had to be sanded, sealed, and finished by hand. The expressions had to be individually painted. In their next film, *Kubo and the Two Strings*, produced in 2016, Laika upgraded their 3D printing technology to use material jetting. This allowed them to 3D print in multiple materials and multiple colours within each print. This removed much of the need for hand-finishing. By this time, the in-house expertise in computer modelling, as well as the use of 3D printing, had time to mature. They then branched out into combining visual and special effects to produce a unique style of animation. In 2016, Laika won the Scientific and Technical Oscar Award for its ongoing innovations in 3D printing for animation.

Working with the technology over time, and being sensitive to the impact that technology has on the workforce, allows an organisation to develop its practice alongside the technology rather than being totally disrupted by it. This example illustrates how an evaluation of the value of additive manufacturing in relation to conventional manufacturing is not straightforward. Many companies, when they first consider additive manufacturing, compare the cost of an existing part in production with the same number and same design additively manufactured. In most cases, the part is not redesigned, and there is no added value in making the shift. Also, the cost of the machines against the straightforward replacement of existing production outputs with additively manufactured ones seems too high. However, as the stop-motion examples illustrate, whilst replacing 700 heads for a figure with 3D printed ones may not justify the cost of a machine, replacing the practice of making 700 heads for a character with 20,000 heads, as in *Coraline* by Laika Studios, would justify the cost. Particularly in relation to the added value of the capability for facial expressions, the additional parts provide the film makers with relative ease of editing these digitally, rather than manually. Projecting possible changes of practice is difficult for companies, engineers, and designers. This potential is key for the effective adoption of the technology.

The impact of digital convergence is being felt in this industry too, where 3D printing is part of the suite of digital technologies available, and the different tools work together to create new ways of manufacturing. In 2020, Laika Studios won the Golden Globe for best animated feature for its film *Missing Link*, beating off competition from Disney, DreamWorks, and Pixar. The film featured 106,000 different 3D printed faces. Furthermore, Laika recently invested in AI for the tedious task of rotoscoping, a process where the tracing is made of a live asset as the basis for animation, and postproduction clean-ups are required on the frames themselves. Laika use the Intel Xeon CUP-based infrastructure to increase the throughput of adding visual paint to their animations and are building on the oneAPI programming model to use machine learning for the repetitive task of rotoscoping. Because of the way the heads are built for the puppets used in *Missing Link*, with changeable 3D printed parts, there are inevitably seam lines that need to be digitally removed. For a film such as *Missing Link*, with approximately 134,000 frames, each photographed individually, anything that can speed up the process helps.

Large-format printing

The innovative artisans at Studio Kite on the Gold Coast in Australia have built a large-format 3D printer for oversize models, suitable for use in the film industry, in advertising, and in public spaces such as museums and galleries. The oversize 3D printer is a propriety filament printer with three heads, called CADzilla, developed by Studio Kite. Where necessary, the models are 3D printed in sections, then assembled to create oversized forms (such as the 18-metre ship the studio produced for the 2022 movie *Thor: Love and Thunder*, which was transported in sections before being attached to a steel frame). Plaster is used to disguise the joins, and for additional detail as required. The scale of the 3D prints designed and made by Studio Kite is quite astonishing (Figure 6.25).

3D printing for costume design

There are numerous, very interesting developments in 3D printing relevant to the design of costumes for the film industry. The most significant of these have emerged from the fashion industry, through the work of designers such as Julia Koerner and Iris van Herpen [13]. Van Herpen's work was influential in shifting the focus in additive manufacturing from engineering to creative practice, helping to bring it into the realm of the product designer. Van Herpen exploits a range of digital fabrication techniques, from laser cutting to 3D printing, illustrated in her unique collections. 3D printing allows for the fabrication of unusual, complex geometries, and new ways of working, as illustrated in Figure 6.26, with interlocking rings as a textile, and 3D printing directly onto fabric. The latter is a straightforward activity. Simply glue fabric onto the base plate and adjust the height of the nozzle to account for the added thickness. Because the print head is heated, using a mesh melts and melds the filament onto the textile.

One of the reasons the special effects industry has been slow to adopt the technology into artistic costume design on the back of the work in functional

6.25
Large format 3D printing by Studio Kite: Chariot for the movie *Thor: Love and Thunder*; statue and column for *Thor: Love and Thunder* prior to finishing (Image courtesy of Studio Kite).

6.26
Textile assembly printed as a single part (a), 3D printing directly on fabric (b).

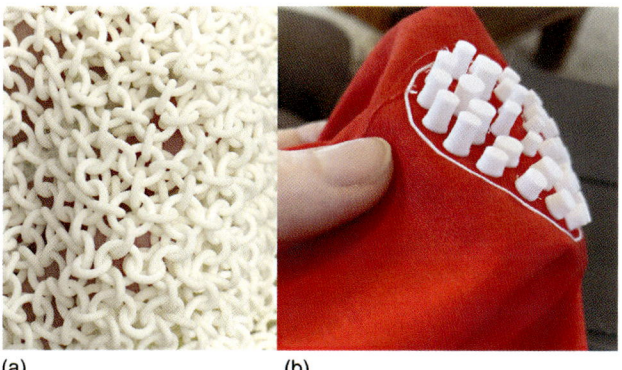

(a)　　　　　　　　　(b)

costume adaptation is because it does require these new ways of working – new thinking in terms of design for additive manufacturing, skills, new work-flow, and a considerable investment in upskilling and in the technology itself.

CASE STUDY 5: 3D PRINTING PERSONALISED DESIGN FOR HEALTH AND WELL-BEING

Technology-enabled healthcare was turbo-charged by the COVID-19 pandemic, and product design futures will be operating at the confluence of technologies that support data-driven, integrated product systems for personalised health care within smart cities, smart homes, and telehealth. The rise of digital com-munication, ubiquitous computing, data analytics, and bespoke digital fabrica-tion (3D printing) enables the evolution of digital-era healthcare tools and practice. Whilst the technical is integral to these developments, the personal must not be lost in the process. In healthcare, 3D printing has predominantly been the domain of engineers and clinical researchers. Product designers have always had to be proactive if they want to influence the design of healthcare products, and the use of 3D printing in healthcare design is no different. Medical development teams rarely involve a fashion designer, for example, even when a product will essentially be "worn," such as with assistive technology. Designers need to be actively involved to provide direction for the essential cultural changes needed to effectively integrate human-centred design and co-creation into healthcare product in a digital era.

Additive manufacturing has been increasingly used for prosthetics, pros-thetic fairings, and customised orthoses over the last decade [14]. Shoe orthot-ics are the most common and accessible applications, but as the technology becomes better understood, more assistive technology applications are emerg-ing. Whilst companies providing these as an accessible service are still rare, there are examples of emerging digital practice that suggest sophisticated sys-tems are being developed. Andiamo in the UK, for example, is a company that was set up by Samiya and Naveed Parvex, parents of Diamo Parvex, who had cerebral palsy. They specialise in the use of 3D scanning as the basis for 3D computer models used for the 3D printing of personalised braces for children with cerebral palsy, which can be changed as they grow. The system integrates big data, machine learning, and 3D printing.

The highest-profile use of additive manufacturing for personalised health-care products has been for low-cost prosthetics for children through the work at e-NABLE [15]. The mechanisms in this prosthetic allow for only a limited range of movements, but, from a design for well-being perspective, the ability for a child to participate in everyday activities, particularly play, as well as the ability for the child to influence the aesthetics of the product add considerable value. The novelty of this project is that the prostheses are mostly designed to be 3D printed on low-cost desktop FFF machines, from the most widely available materials. This makes them extremely low cost compared to comparable commercially available prostheses, and open opportunities for anyone with a 3D printer to fabricate them and donate them to those in need.

Meanwhile, the Exo prosthetic by William Root and the work of Bespoke Innovations, Inc. demonstrate how the technology enables designers to shift the product from purely functional towards a well-being design. In this company, an industrial designer and an orthopaedic surgeon aimed to develop individual prosthetics and fairings for their clients to meet emotional as well as functional needs – for example, in designing a leg prosthetic for a footballer or a cyclist, which are different [16]. This change in thinking allows designs to be perceived as enhancements, rather than purely functional replacements. 3D printing is well placed to support this shift in perspective not only for prosthetics, but for all assistive technology. Arguably, if these types of orthoses continue to develop, they could be used to support balance and movement as people age, enabling them to remain independent for longer. Again, the shift in practice is more about changing cultural acceptance of these types of products, and subsequent development of suitable products and accessible systems, than about any technical constraints.

Another advantage of shifting from mass production to customisation with 3D printing is the digitisation of previously manual processes. In orthopaedic clinics, for example, occupational therapists currently tend to build splits by hand, manipulated into the correct shape for an individual from flat sheets of plastic, informed by their own expertise. This is a relatively slow process, and one that cannot be mechanised or tracked. Therefore, it is highly dependent on the individual skill and experience level of the therapist. If a systems approach can be introduced, one that may combine 3D scanning and some level of automated CAD workflow, it could improve consistency with knowledge built into the system by the practitioners, rather than lost if they move on. This is certainly being explored in applications including immobilisation devices for radiotherapy [17] or the production of artificial eyes and orbital prostheses [18].

As an example, Orthokids in Australia uses 3D printing to produce infant helmets to help correct skull development issues, based on patient scans. Not only does this provide a systems approach, but it also removes the need for the baby to be subjected to plaster mould construction, which is a very difficult process for infants to be involved in (and their parents). The 3D CAD model can also be used as the basis for evolving the model as the child grows and the problem is corrected. This iterative development illustrates the potential of 3D printing in healthcare.

The influential work of design researcher Graham Pullin [19] on design for ability has highlighted the need for the language of assistive technology to shift from the medical to the wearable. He uses the example of spectacles development and its transition from a medical product to a fashion one to illustrate how this can happen. 3D printing is enabling design for health products that are bespoke to the personality of the individual, as well as their practical needs.

Assistive technology is beginning to evolve towards user-centred design as defined by product designers, rather than technical specialists and clinical practitioners.

Custom splints

The SCAN2CAST project in Figure 6.27 began as an exploration of a novel CAD modelling approach called "implicit modelling" in the development of a systems approach to mass customisation. It provided a platform for the specification of a product from an acceptable range of parameters that were established for that platform. Implicit modelling refers to a light-weighting method of representing complex 3D objects using mathematical functions to describe solid bodies, making it adaptable to computational design, which is also formulae-driven. This makes it suitable for mass customisation, which is also, typically, formulae-driven. The advantage of implicit modelling is that the CAD file sizes remain significantly smaller than the more conventional representation of complex forms using meshes or patterning features within solid modelling software. This makes it far quicker and easier to design and modify a model without relying on high-performance computers with expensive graphics cards.

One of the strengths of computational design in the context of additive manufacturing is its ability to enable the creation of complex lattice structures that conform to complex organic surfaces. This makes it suitable for the creation of custom fitting casts, splints, orthoses, or prosthetic sockets. Splints for physiotherapy applications were chosen in this project as they provide for a good "low-hanging-fruit" application that does not, typically, deal with the complex medical parameters of prosthetics for amputees, or medical casts (e.g., dealing with swelling, the time between the break and the need for the cast, etc.).

Several solutions are currently offered in the 3D printed splint market (ActivArmor, Xkelet, Spentsys, etc.), mostly with custom written software

6.27
Examples of SCAN2CAST 3D printed custom splints (Diegel).

applications. In discussion with local physiotherapists, many of whom work alone or in very small practices that do not have "technicians" or software experts, one of the challenges with the existing software solutions was that they required the user to develop a high level of CAD skills. This typically involved sketching in 3D on the scanned model of the arm to indicate where they wanted trim lines or splits to be, and operations for smoothing certain areas. Many of the physiotherapists spoken to indicated that they preferred a solution in which they did not have to develop new CAD skills that were outside of their area of expertise.

For these reasons, the SCAN2CAST system was developed in which the only work the physiotherapists had to do was move three straight lines for the trim lines (indicating the top and bottom of the splint as well as the thumb exit end of the splint), and two more straight lines for the split lines (to break the splint into two separate "clamshell" components). If greater accuracy was required, such as a higher degree of curve than in the straight split lines, they could integrate this into the CAD model, but this was not a required skill to drive the splint generation system. In the SCAN2CAST system, the physiotherapist imports the 3D scan of the arm, indicates the three trim lines and the two split lines, and chooses the thickness of the splint, depending on the level of flexibility or rigidity they require, and the details are automatically generated by the software.

From just the trim line and split line information, the splint can be automatically generated as shown in Figure 6.28, with a Voronoi-type lattice as part of

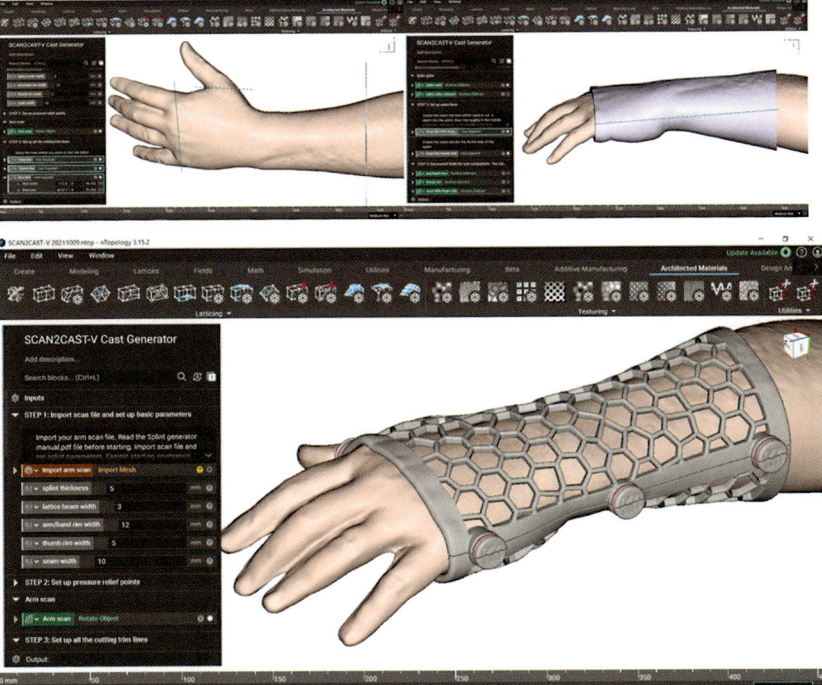

6.28
Setting up the trim and split lines to automatically generate a ready-to-print splint.

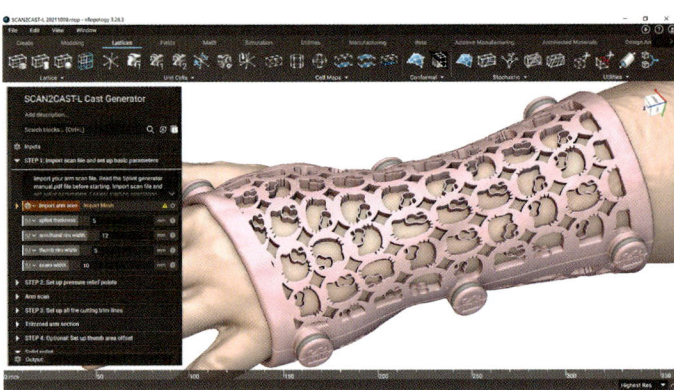

Figure 6.29 Custom-made, 3D printed decorative splint (Diegel).

the design and be ready to print. However, to offer further flexibility, the physio-therapist also can relieve or increase pressure in certain areas. This facility is needed where these is an exposed sore, for example. By moving the spheres provided in the model, the physiotherapist can change the pressure and input in a specific zone to increase or relieve the pressure. Once a workable solution was established, the aesthetic aspects of mass customisation could be taken further to allow the physiotherapist or customer to aesthetically adapt the splint to suit their style. To do this, a workflow was developed in which instead of a default Voronoi-type lattice, the physiotherapist could select a lattice cell from a library of pre-sets or import their own unique design, as in Figure 6.29. While this might require some minimal CAD skills to adapt, this allows for a wide range of customised splints to be available to the user.

When printed using polymer powder bed fusion (SLS or MJF), the splints are rugged and durable, and they are dishwasher safe so they can easily be kept clean. They are also light and breathable, so there will be a reduction of sweat on the arm and the wearer can access the skin to scratch any itches that might occur. To secure the splints to the arm, ideally by the person wearing the splint when on their own, various attachment systems were investigated and tested, including rubber bands, O-rings, elastic cords and cord clamps, hinges, and locks. To keep costs low, and because they were easiest to put on unassisted, a moulded elastic ring with a handle was chosen as the final attachment system as a good compromise between ease of use and economy.

A range of the splints was 3D printed and tested for extended periods to test for comfort, durability, and ease of use. The products performed effectively, and the system provided a cost-effective solution to producing splints and casts with enhanced comfort and connection for the user.

Changing Perceptions

The beautiful exploratory work of product designer Charlotte Dickson, shown in Figure 6.30, was developed as part of the health and well-being design research led by Professor Steve Reay in New Zealand. It illustrates the opportunities

6.30
Stigma-to-Silver-
Linings (designed
by Charlotte
Dickson).

additive manufacturing provides to product designers in working on bespoke healthcare designs. Dickson's master's work investigated tracheostomy products and developed designs that would help to reduce the discomfort and stigma she found associated with those living with the effects of the procedure [20]. In a tracheostomy, a breathing tube is inserted into the trachea/windpipe to bypass a blockage, such as that caused by a tumour. Whilst some patients need the tube for only a matter of weeks, others rely on it for many years. Dickson found that the tubes themselves had changed little in the 140 years since they were invented, and yet patients found the product both uncomfortable and unsightly, as well as difficult to clean. For long-term users, they reported that the product affected their sense of self, and their subsequent ability to form social attachments. Many users reported trying to modify the product, usually by changing the provided strap in the first instance, to make it less unsightly.

Through co-design activities, Dickson developed a brief for a product that improved both the aesthetics and usability of the product. She was keen to change the perception of the product, moving it from an overtly medical one to a product by which the patient could feel empowered. 3D printing was a critical enabler for the project because it allowed her to customise the products to an individual's unique anatomy and was a viable product to produce in metal. To re-enforce the message of the product, Dickson houses it in a CNC carved American Ash box, shaped to fit the bespoke components. Whilst this product is still awaiting clinical trials, the project illustrates how the technology can contribute to building user-sensitive products to support health and well-being.

Prosthetics with Personality

Product designer Eloise Cleary was equally motivated by the well-being of the wearer of healthcare products (Figure 6.31). Her major project in the final year of her degree was a project called Reboot, which used 3D printing to create an

adjustable orthopaedic ankle boot for rehabilitation. Her subsequent honours work focussed on the redesign of prosthetics enabled by 3D printing technology to support the mental health of users whilst meeting changing functional requirements. Cleary wanted to co-create a foot prosthetic that was an extension of the wearer's body and personality. The result was a product called Super-abled, depicted in Figure 6.31, which features a 3D printed metal base with polymer 3D printed, interchangeable covers, moving from practical to extravagant. The base has a height-adjustable heel, which Cleary designed to allow the user to easily shift from casual day mode to evening wear.

In both cases, the driving factor was not 3D printing technology, but user-centred design and co-creation. The projects were enabled by the technology both in terms of the ability to create viable outcomes and the creative thinking it allowed.

In 2017, Lionel Dean of FutureFactories Studio developed an illustrative example of functional, human-centred design of a lower-leg prosthetic enabled by additive manufacturing, shown in Figure 6.32. Dean's design process to create the Angel Leg used storytelling as a way of forming a narrative for the wearer, weaving in detail personal to her life. He then built the design from the narrative. This is very characteristic of his work. The leg was 3D printed in titanium on a powder bed, selective laser melting (SLM) printer. Dean has demonstrated numerous innovative ways of connecting with the user to maximise the bespoke 3D printing the technology enables. This includes the development of storylines for his products that help engage the user, such as in his gold heart jewellery series, featured on his FutureFactories Studio website.

6.31
Super-abled
(designed by Eloise Cleary).

6.32
Angel Leg (Lionel T.
Dean,
FutureFactories
Studio).

CASE STUDY 6: ADDITIVE MANUFACTURING MEDICAL DEVICES

The use of 3D printing in medical applications has a dedicated field of research and industrial suppliers. From a product design point of view, much of this work is situated too far into biomedical engineering to be accessible, or of interest. However, as 3D CAD modelling and design for additive manufacturing becomes more prevalent in medical applications, and a greater range of 3D printed materials become validated for use, designers do have highly valued skills even in highly technical areas, such as around bone regeneration work and biofabrication. It will be interesting to see over the next decade whether clinicians learn more 3D CAD and design for additive manufacturing as part of their training, or product designers and engineers learn to work in more medical contexts.

It is likely that, in the future, there will be a greater integration of practice between designers and clinicians using 3D printing, especially where the technology enables greater communication between patients and clinicians, and even between clinicians with different specialisations. Product designers are being employed in clinical research teams for their abilities in design for additive manufacturing, even where there is a predominance of engineers, as they are trained to understand the people in the process as well as the technology.

3D printing communication tools
At its most fundamental, 3D printing can be used in clinical settings as a communication tool between clinicians and patients, and between surgeons themselves. Some procedures are difficult to visualise, and therefore discuss, so 3D printed models are increasingly used to help break down those communication barriers.

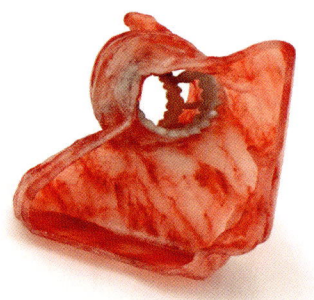

6.33
3D printed artery
model with stent
(Diegel).

Designers and engineers play a key role in creating these models, typically from patient scans such as CT (Computed Tomography) or MRI (Magnetic Resonance Imaging). These scans are made up of many 2D cross sections and require advanced software to convert these into 3D models, in a process known as segmentation. Depending on the complexity of the anatomy and/or the desired model outcome, design-based processes may be needed to refine the model (e.g., smoothing), isolate specific anatomy or pathology, and prepare the model for a suitable 3D print process. For these reasons, designers and engineers play a vital role in this process and are well equipped for working with the advanced CAD software required.

The artery with a stent model, shown in Figure 6.33, is used by doctors to explain to patients what will be happening to them during their operation. 3D printed medical models are becoming popular for multiple purposes, including education, surgical planning, and even pre-surgery practice.

Prototyping a novel medical implant

For start-up companies, speed and agility are key to getting a new product to market. This is particularly true in the highly competitive medical device industry, which is dominated by large global brands. For start-up medical device manufacturer MAXONIQ, collaboration with a multidisciplinary team at the Herston Biofabrication Institute (HBI) allowed an idea to rapidly evolve from concept to provisional patent in just six months because of 3D printing. The concept the team came up with was for magnets to be implanted in patients as a means of retaining oral and facial prostheses. This would avoid the complications associated with bone mounted posts, which ordinarily protrude through the skin and require constant cleaning, or glues or other mechanical methods of retaining prostheses. The magnets would ultimately be embedded within a sealed titanium housing to make them safe within the body. However, working with titanium is expensive and made more complicated by the small scale of the implant envisaged. This would not be suitable for rapid iteration.

FFF printing with PLA material meant that dozens of different ideas could be 3D printed in less than 30 minutes on a standard desktop machine. Small 3×3 mm and 5×5 mm neodymium magnets were glued into the printed housings and tested on a 3D printed skull, including locations for ears, orbits, and

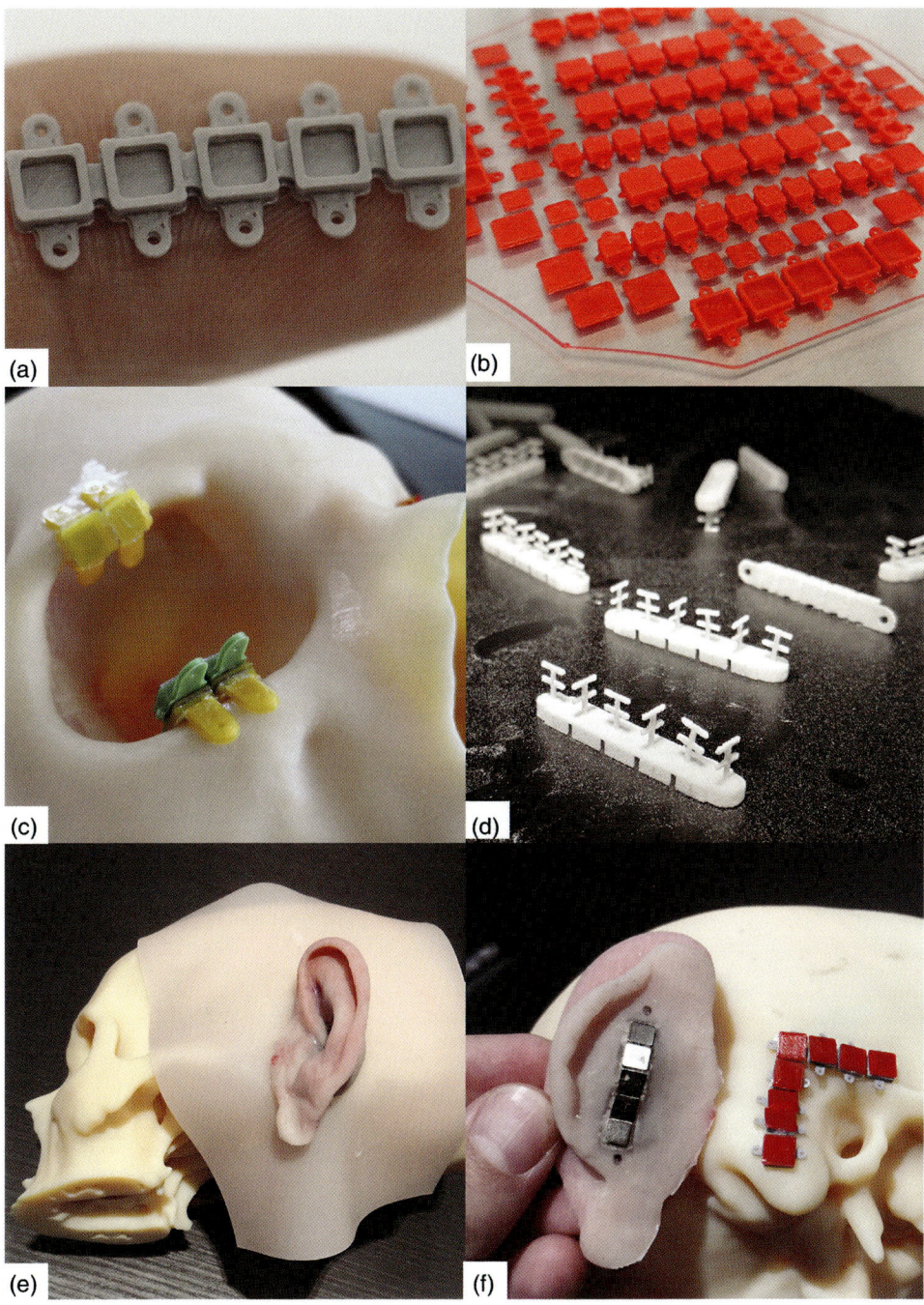

6.34
Detailed view of a prototype implant housing which was printed using FFF (a); a collection of different prototypes can be printed with FFF in <30 minutes at this scale (b); testing the fit of different-sized magnet housing on different 3D printed skull anatomy (c); a collection of SLS printed housings to be embedded in the silicone prosthesis (d); fitting of a prosthetic ear using the magnet system over a skin-like silicone sheet (e); detailed view of early experiments directly gluing magnets into a prosthesis to match the 3D printed implant design (f) (Novak).

noses. Different silicone sheets could replicate skin, and the team performed rough mechanical tests to understand how magnets with different strength, as well as different arrangements of magnets, affected retention. Changes could be quickly accommodated by the designer and a new batch of prototypes produced. SLS was also useful for testing geometries that were not well suited to FFF. For example, the team experimented with different features to help the corresponding magnet housing to be embedded in the prosthesis and ensure it remains within the silicone when being pulled off the implanted magnets. Small posts and spikes with <1 mm thickness could be fabricated with SLS due to there being no need for support material, which could not be done with FFF.

3D printing in this case enabled the concept to mature rapidly and reach an important go/no go decision point. Based on the 3D printed prototypes, MAXONIQ and the HBI team had confidence to secure a patent and engage a magnet manufacturer to fabricate the titanium-encased magnetic implant based on the FFF design, suitable for running a small clinical trial. It is estimated that the total costs of 3D printing to reach this point was less than US$350/300 pounds sterling, which is an incredibly small materials budget to progress from concept to an initial production run (Figure 6.34).

Anatomical training model for neurosurgery

3D printing is increasingly being used to create education and simulation models of human anatomy for training doctors and surgeons. Evidence suggests that for many procedures, 3D printed models perform as well as animal models or cadavers, yet can be significantly cheaper and more accessible [21]. There are also numerous benefits over the traditional "see one, do one, teach one" approach to learning clinically relevant skills, allowing clinicians to gain confidence and experience prior to operating on a patient.

As an example, neurosurgery is a specialist medical discipline dealing particularly with the brain, spinal cord, and peripheral nerves. An External Ventricular Drain (EVD) procedure is one of the most common procedures performed by neurosurgeons and involves passing a catheter through the brain and into the ventricles. This may be needed after severe head injury or in patients suffering from hydrocephalus to reduce excess fluid and relieve pressure. Accurately locating the ventricles is critical to minimising damage to the brain. Industrial design student Liam Georgeson developed an advanced training model for performing an EVD as part of an internship at the Herston Biofabrication Institute. This involved early experiments with a Stratasys J750 Digital Anatomy Printer (DAP), capable of multi-material 3D printing and producing "parts" (or anatomies) in materials with similar mechanical performance to bones, blood vessels, organs, and numerous soft tissues all in the same print. Regular feedback from neurosurgeons informed iterative development.

However, rather than design a complete head to be manufactured for each trainee of a workshop, which would be particularly expensive if completely 3D printed on the J750 DAP, Liam developed a modular system (Figure 6.35). This

6.35
Experimenting with materials and simple prototypes from a Stratasys J750 Digital Anatomy Printer to gain feedback from neurosurgeons about haptic and other qualities compared with real anatomy (Novak).

meant that only a small cartridge would be 3D printed and replaced each time. An overall head model, needed to be 3D printed only once or in small quantities for running group workshops, could use lower-cost technologies like binder jetting or selective laser sintering. The external casing of the cartridge was 3D printed using FFF to further minimise costs. Only the internal "cylinder" was made up of numerous layers of 3D printed materials representing scalp, periosteum, skull, dura mater, brain, and ventricles, costing approximately US$85/72 pounds sterling in J750 DAP materials.

In Figure 6.36, the left image shows a prototype of the cartridge system, with the left "cylinder" being 3D printed on a Stratasys J750 Digital Anatomy Printer, before being placed in a FFF casing. The right image shows a neurosurgeon testing the 3D printed dura and brain materials after performing a burr hole, providing feedback to inform material choices and other design features during development.

6.36
Cartridge system models designed by Liam Georgeson.

In medical training applications such as in Figure 6.37, 3D printing provides new opportunities to educate clinicians, increasing opportunities for hands-on skill acquisition. New industries are forming around the design and 3D printing

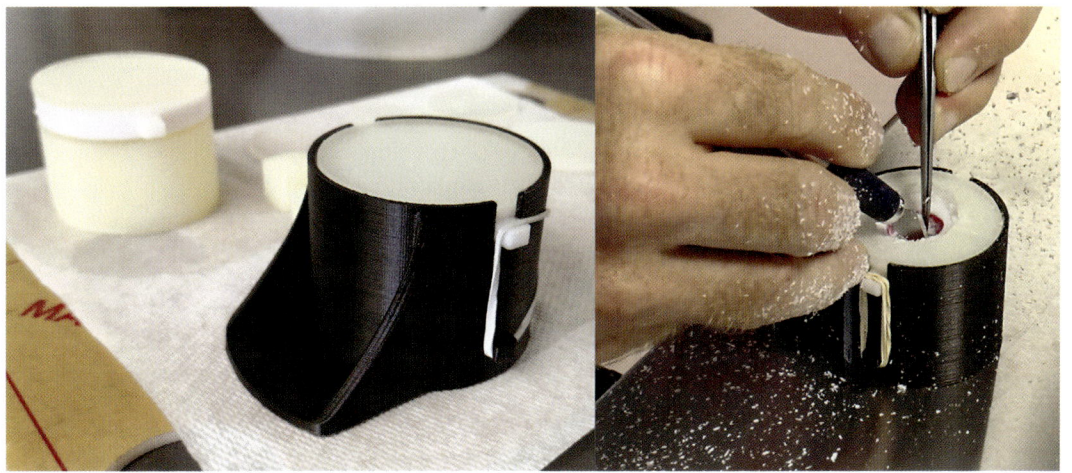

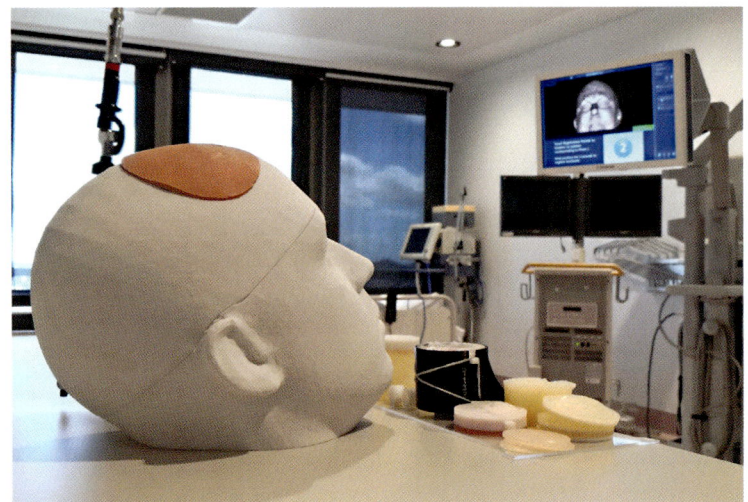

of such models, which can be generic for many applications, or even patient-specific to enable pre-surgical planning for complex cases. Designers are playing an increasingly important role in this emerging industry.

Optimisation of a radiotherapy immobilisation face mask

During radiation therapy, it is important to immobilise the area of the patient being treated for cancer. This minimises damage to the healthy cells surrounding the cancer site. When the treatment region is the head, the use of a thermoformed polymer face mask has been the gold standard over recent years. This is custom formed over a patient's face; however, the experience can be stressful and claustrophobic as the warm plastic sheet is stretched and pressed over their face. There is also the risk of the mask not being correctly fitted, allowing unwanted movement. Further, it is highly dependent on the skill and experience of the person fitting the mask, and the person on whose head it is being fitted, which may change over time, for example, if the patient loses weight.

3D printing has been explored in this context in an increasing number of studies [22], and 3D printers are increasingly available within hospitals. With the aid of 3D scanning, a patient's face, head, and neck can now be digitised without any contact, and in a matter of seconds. A team of researchers at Deakin University and the Peter MacCallum Cancer Centre have been developing the workflow to fabricate patient-specific immobilisation masks, as well as exploring several opportunities offered by 3D printing. In one series of designs by PhD researcher Amirhossein Asfia, FFF technology was the focus of experiments due to the accessibility to this type of 3D printing directly within the Peter MacCallum Cancer Centre. 3D scans using a handheld structured light 3D scanner from Artec were the first part of the process, before the desired region was thickened to create a "blank" mask. ANSYS software was then used to topology optimise the mask [23]. This reduced material in regions that did not add to the

6.38
Topology-optimised
mask produced
using FFF being fit
tested to a phantom
(a, designed by
Amirhossein Asfia);
full-colour MJF
printed mask
featuring
lightweight lattice
structure and
markers for tracking
(b, designed by
Faizan Badar).

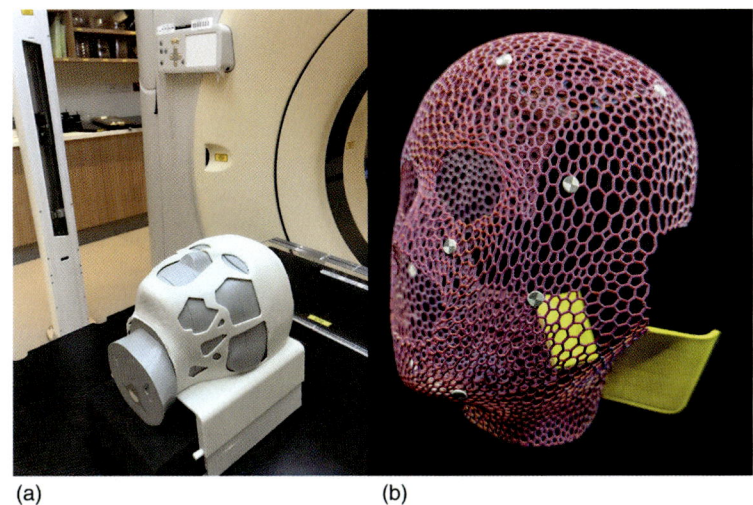

(a)　　　　　　　　　　　　　(b)

mechanical properties of the mask under expected loading conditions, as well as considered some of the limitations inherent with FFF 3D printing such as the need for support material. The final prototype can be seen fitted to a phantom and fixed to a treatment couch (Figure 6.38a).

In another design by PhD researcher Faizan Badar, MJF 3D print technology was the focus of prototyping to further reduce material use and explore opportunities for full-colour 3D printing (Figure 6.38b). Manipulation of the 3D scanned mesh allowed Faizan to control the density of the resulting lightweight structure that utilised a Voronoi pattern. This type of structure can be automatically generated from a mesh file in software such as Autodesk Meshmixer. Colour was also used to improve the aesthetic qualities of the mask, which could be particularly important for children. Further, it enables one to propose how additional features, such as markers, could be integrated into the design to allow for real-time tracking of patient movement or aid in positioning for treatment.

3D printing a reusable N95 respirator mask

At the height of the worldwide COVID-19 pandemic, usual supply lines of personal protective equipment (PPE) were threatened due to increased demand and disruptions in the manufacturing and logistics supply chains. A multidisciplinary team of clinicians, biomedical engineers, and industrial designers co-designed and evaluated a reusable respirator mask [24]. Specifically, the companies and groups involved were Metro North Health, Herston Biofabrication Institute, Royal Brisbane and Women's Hospital, Digital Metro North, and Shapelabs.

An N95 mask is a respiratory protective device designed to filter 95 percent of airborne particles. Respirator masks protect the wearer against inhalation of airborne particles greater than 0.3 μm in size. Following design-based workshops between stakeholders in March 2020, 3D printing was used to rapidly test ideas. There were two major challenges to address in the mask design:

(i) creating a comfortable interface with the face while providing a reliable seal, and (ii) accommodating a no-leak filter housing.

In total, six rounds of iterative prototyping were completed, as well as 26 fit tests with a broad diversity of face shapes, nine gas exchange tests, two simulations, and two comfort tests. The result was a modular design available in three sizes (small, medium, and large), composed of:

- A flexible face seal (vacuum-moulded silicone).
- Snap fit, rigid inner and outer shells which feature four pegs to attach two straps of buttonhole elastic or gag-tie, like those used to secure face shields on the head. The shells were 3D printed using the Herston Biofabrication Institute's selective laser sintering 3D Systems sPro 60 in Duraform PA biocompatible nylon.
- Three holes that can accommodate a commercial Heat Moisture Exchange Filter (HMEF, with a filtration level ≥ 99.97 percent for particles 0.3 μm in size or larger), a 3D printed filter casing with a cut-out of N95 paper, or a 3D printed nylon plug.

In the project shown in Figure 6.39, the ability to produce the prototypes locally at point-of-care allowed for rapid design feedback and device evaluation. 3D printers available at the Herston Biofabrication Institute were diverted from other projects to accommodate the production of prototypes and a final design. The final N95 mask was designed to be 3D printed using SLS technology as a means of circumventing the supply chain challenges of sourcing standard N95 masks or attempting to get them mass-manufactured based on this design. SLS printing was available at the Herston Biofabrication Institute, directly within the hospital precinct, and allowed for rapid supply to the hospital.

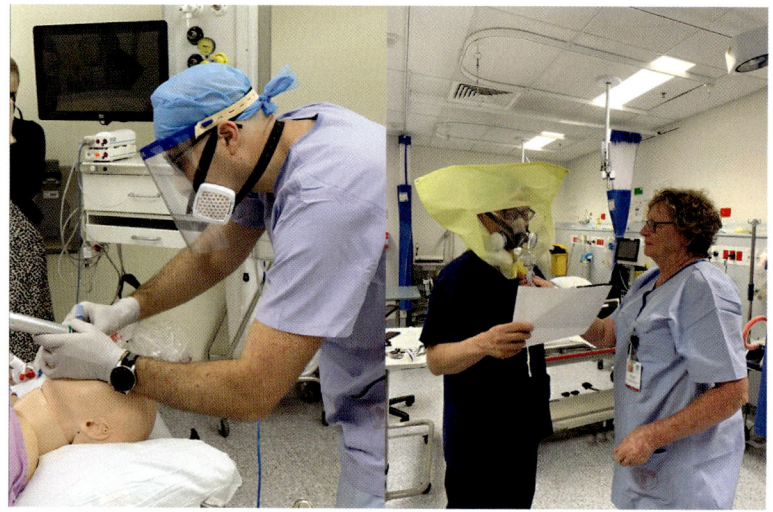

6.39
Intubation simulation with the mask fitted with square filter housings (left), fit testing with the Halyard qualitative fit testing kit (right). (Images courtesy of Mathilde Desselle, Herston Biofabrication Institute).

CASE STUDY 7: CREATIVE EXPLORATION: PLAYING AT THE BOUNDARIES

Whilst commercial outcomes are the basis of the product design profession, exploring the geometrical complexities enabled by 3D printing both for the fun of it and to create new ways of seeing, enables the investigation of new avenues for practice to inform design, outside the engineering and clinical domains. Experimental art, and designing for less functional products such as jewellery, allow designers to push the boundaries of what can be designed. 3D printing in art has several high-profile practitioners, such as Barry X Ball, Michael Eden, Richard Dupont, and Anish Kapoor. Their work utilizes the unique opportunities 3D printing brings but integrates into their practice, expressing concepts and commentary. The strange, voluminous sculptural work of Belgium artist Nick Ervinck, and the spike-backed, "chairs in motion" Meteor chairs by MAD Architects, exhibited at the Milan Furniture Fair, illustrate the very different directions that this exploration can go.

There is also inspiration from tech-focused fashion designers using 3D printing, such as the inimitable Anouk Wipprecht, whose startling designs challenge perceptions of what is or could be. Wipprecht creates responsive designs, such as in the 3D printed Spider Dress, which has animatronic limbs and embedded sensors that react to proximity. For product designers building an understanding of what is possible with 3D printing and how to work with it, a good place to start is creating samples of different textures and geometries, illustrated in Figure 6.40. 3D printing enables assemblies as a single part, which is a good basis for textiles, and textures on the prints themselves, which can then inform design specification for products where texturing can be functional, such as for grip or aerodynamics.

The other constraint that is frequently brought up in discussion with engineers, and on occasion with product designers, is concern that established disciplinary software is not conducive to creating the complex or organic forms that are capable of being 3D printed. The need for plug-ins to conventional product design 3D modelling software does speak to a lack of development in dedicated product design programs in this realm. Architects are focused on robot arm

6.40
Creating samples as a catalogue of digital printing to inform practice and for communication with clients.

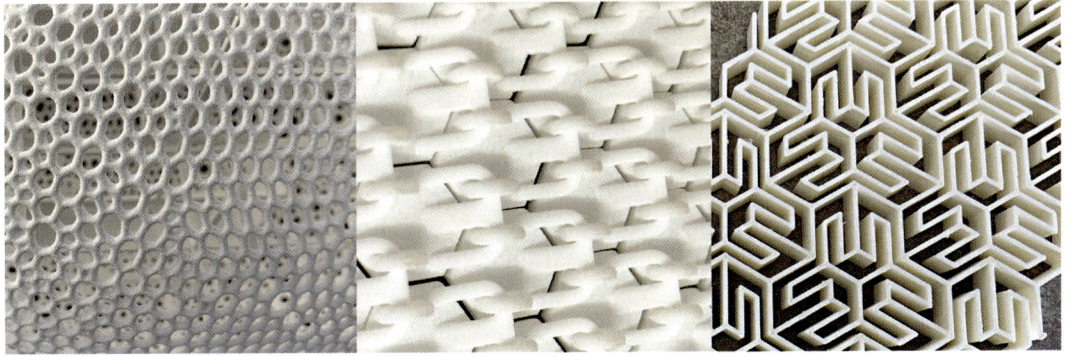

6.41
Exploring the ability to create form with 3D printing using SLM, SLS (right), and MJF (left), designs modelled in Solidworks (Loy).

printing, with the Grasshopper plug-in for Rhino featuring in much of the work [25]. For product designers, 3D modelling software does allow for relational modelling and for working in the 3D environment, both appropriate for the bespoke capabilities of the printing technology, but integrated organic, 3D modelling within the domain still needs development. However, it is still possible to create organic forms within conventional solid modelling programs, and increasingly product designers are venturing into the use of software more commonly associated with the film industry. It will be interesting to see where this goes. In the meantime, the use of additional planes, inserted at different angles in the 3D space, drawing in 3D space, and the creation of pathways within the model as the basis for geometry, allows for an exploration of creative forms, such as in Figure 6.41.

Product designers, as well as designer-makers who have worked across programs unconventional to the discipline for some time, have shown how to draw organic forms in 3D that illustrate the full potential of 3D printing in this context. Product designer and artist David Haggerty demonstrated this in his early work using dual-filament 3D printing (ABS and PLA), in Figure 6.42. The design is an excellent example of organic modelling using Solidworks.

Haggerty went on to explore the integration of digital making with traditional practice, which is an interest of many designer-makers. He designed a range of jewellery called Fractilus. In this work, he designed geometrically complex pieces to demonstrate the potential of 3D printing. He then worked with the online service provider iMaterialise to produce direct-to-manufacture 3D prints in steel in a powder bed, laser printer (as in the first example in Figure 6.43) and in wax printed forms using stereolithography to produce a wax-like resin that were then translated into end-use parts through lost wax casting (as in the bronze example with PU coating in Figure 6.43). Unfilled, paraffin-based wax can also be used with material-jetting processes. For an entrepreneur, the ability to work with an online service provider opens the potential of global

6.42
Exploration of organic form, designed, 3D modelled, and 3D printed by David Haggerty.

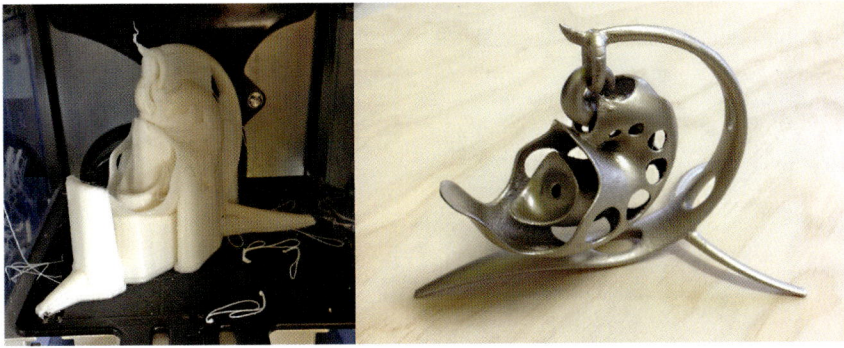

6.43
Fractilus range of jewellery designed by David Haggerty, using 3D printing in different ways during the process direct metal printing (a), and lost wax casting (b).

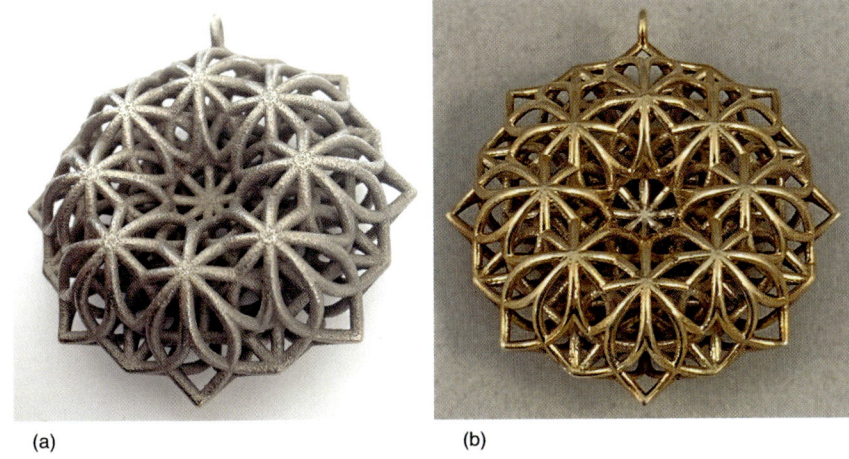

(a)   (b)

markets and remote working. After the pandemic, this could be of increasing interest to product designers around the world.

The design by Haggerty in Figure 6.43 is deliberately at the boundaries of geometrical complexity possible with the technology. Both pieces were a challenge in post processing for iMaterialise because the finishing process involves tumbling with corn polishing media, with finishing details completed by hand.

Exploring the boundaries of making, the work of Benjamin Dillenburger and Michael Hansmeyer in collaboration with 3DXL combines architecture and art. Emphasising the experience created by their work such as the "arabesque wall" (following Hansmeyer's earlier work, Digital Grotesque, ETH Zurich), the piece was created using sand 3D printing in an innovative application as the technology is generally used in industry for producing moulds. The innovative work of New Zealand industrial designer Nicole Hone, in collaboration with Ross Stevens as part of the MADE group at Victoria University of Wellington, called Polyphytes, uses the design software Houdini to create exquisite designs with material jetting 3D printing to explore the physical effects of different substances flowing through a series of fine capillaries, mimicking the vascular systems in nature,

6.44
Material jetting vases, showing prior to post-processing through finishing, and a large example of a print (approx. height 600 mm and a close-up showing internal detail) (Diegel).

such as in leaves. Translucency and flexibility in the process, as well as the ability to produce high-colour outputs, were critical in this work. The vases, shown in Figure 6.44., are also created using a form of material jetting. The images show how the part comes off the machine prior to finishing, and how it looks after polishing.

Captured parts, hollow parts, and assembly as a single part

Solid metal 3D prints are straightforward if they are small scale; anything large will be more problematic because of the build-up of heat within the object, potentially warping the part. Support structures must be designed into the print. The challenge with support structures in any of the processes that require them is that removing them can damage the surface of the part, and this needs to be designed for and then addressed in post-processing – much as parting lines may need to be finished out of an injection-moulded product. However, beyond solid parts, the ability to create mechanisms as a single part is a critical advantage of the technology.

For product designers who work with existing manufacturing technologies, the idea of creating products with captured parts is a radical diversion from standard practice. In 3D printing, particularly with powder bed technologies, because the 3D model is sliced into layers and each layer printed as a series of lines and points, captured parts are straightforward to create. In the example in Figure 6.46, the ball within a ball was 3D printed using MJF.

However, the earrings shown in Figure 6.45 had to be made with a hole in the back of each one to release the trapped powder. A recess was designed into the Meshmixer-modified, open-sourced 3D scans, and off-the-shelf earring studs epoxied into the holes. The aim was to make the earrings as light as possible, so they do not weigh on the earlobe. The intent of these parts was to explore how small a print could be made using a standard powder bed, aluminium laser printer, and still retain detail in the output, illustrated in Figure 6.46a. A slightly larger version retained noticeably clear detail, as shown in Figure 6.46b, but would be too heavy as jewellery at approximately 30 mm high. Nevertheless, both versions demonstrate the standard of detail that can be produced. The earrings were printed in large batches, covering the base plate.

6.45
Ball within a ball
trophy design
(Novak) and
BLING3D earrings/
cufflinks (Diegel).

6.46
Smallest versions of
the earrings (a) and
larger size showing
the level of detail
that can be achieved
(b) (Diegel).

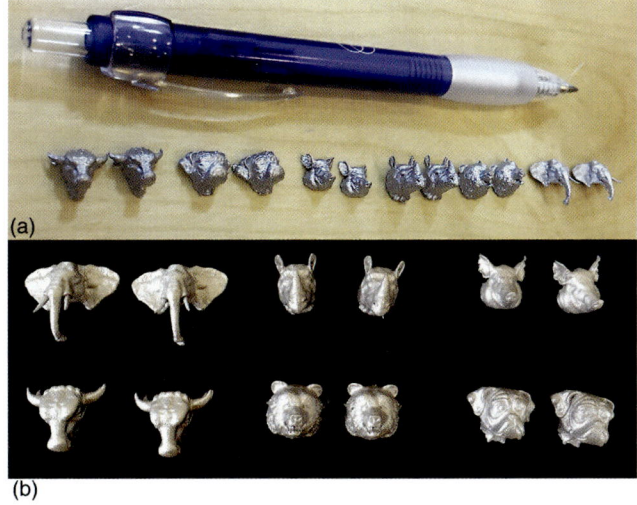

The final example in this section is a series of bracelets, shown in Figure 6.47. This work began as an exploration of an assembly as a single part, using aluminium. This work was started relatively early in the development of design for additive manufacturing when most metal parts were single-piece parts only. Topology optimisation was common at that time, but not assemblies in metal. The first design was used as a way of figuring out what type of mechanism could be printed with moving parts, and what sort of linkages would be needed on the back of the tiles to join them together. And, of course, it had to be printable with no support material.

The designs were printed face down because removing any support material off the mechanism would be too difficult. Whilst assemblies with moving parts are relatively easy to 3D print using polymer formats, working with metal

6.47
Underside of failed experimental print (a); series of postprocessing steps (b); finished bracelet designs (c) (Diegel).

is difficult because of the support structures that need to be attached during the process, and because the parts must be attached to the baseplate. This is further complicated by difficulty in metal powder removal, and surface retrieval in post-processing when using metal than in polymer. This is because metal powder can get embedded in the skin, much like splinters, and sintered metal is more stubborn to remove than nylon. Figure 6.47a shows the underside of the bracelet. The angles for the movable parts in this assembly were determined after numerous experiments to avoid the need for supports, as it would be too difficult to remove supports from the underside of each link. There was considerable experimentation and redesign needed to find the outcome that met the brief, reduced post-processing, and did not fail during printing.

Post-processing also involved experimentation to create a reliable workflow that reduced damage to the parts and the time taken for each step. Removing support from the surface of the tiles was straightforward, then the surface was linished on a belt sander to clean away the roughness left on the surface by the supports. They were then sand blasted for a satin finish. The final step was polishing which took a considerable amount of time by hand, with each tile having to be polished individually. However, in a batch-production workflow, embedding the bracelets in clay or a similar material would allow the tiles across the entire batch to be polished in one go. The four images together, shown in Figure 6.47b, show the post-processing steps as follows:

- Remove from build platform and remove support material: 30 minutes. Tools: Bandsaw and pliers.
- Use a linisher on the face surface of the bracelet to clean and smooth: 30 minutes. Tools: Belt-sander/finisher.
- Media-blast with glass beads: 3 minutes. Tools: Media-blasting cabinet.
- Polish: 4 hours. Tools: Sandpaper, buffing wheel, and polishing compound.

Once the design for hexagon tiles was resolved, the designer branched out into a fish-scaled version and a circular tile, just to confirm that it worked for different configurations. Magnets were used as the clasp mechanism (because of the satisfying click they make when closed).

Product designers tend to be curious. Working at the creative and technical boundaries of the capabilities of a technology, whether additive manufacturing or any other technology, helps progress its development by indicating new directions to explore within more conventional project work. 3D printing is a new world, and one that is far more expansive than has been explored so far. Designers are only at the start of their involvement with the technology, and with designers looking to the future, the use of 3D – and 4D – printing is in its infancy [26].

CASE STUDY 8: CUSTOMISABLE TECHNICAL PRODUCTS

The value of the defined strategies of this book is that they break down what may seem like complex approaches to adopting AM into manageable chunks. Through their combination, however, seemingly conventional products that may not benefit from just one approach can be completely re-thought from a technical perspective. These may not replace conventional products but address the needs of niche markets looking to enhance performance in some way. Such products are often at the intersection of design and engineering, utilising advanced software to generate and evaluate 3D models, as well as high-performance materials and processes.

Open source, customisable surf fins for 3D printing
Surfing is a sport that embraces maker culture, with recreational and pro surfers alike keen to experiment with board and fin shapes, as well as placement and

configuration of fins, to help them catch the perfect wave. 3D printing has digitised this experimentation across the spectrum of surfing sports, including kite-surfing and stand-up paddling.

James Novak has been 3D printing his own kite-surfing fins since 2014, utilising desktop FFF 3D printers to create custom low-cost fins. These can be tested quickly, as a 3D printed fin trialled in the afternoon can be parametrically altered from a few dimensions and 3D printed overnight, ready to test again the next day. Novak has also freely shared designs like this online through 3D file sharing websites such as Thingiverse, as well as his blog (https://edditive blog.wordpress.com/category/kitesurfing-and-sup/). A Creative Commons license (Creative Commons – Non-Commercial – Share Alike, CC BY-SA 3.0) allows anyone around the world to download, copy, 3D print, and "remix" the design to suit their needs. This is a new way for surfers to connect and share ideas. It provides those without traditional hand-shaping skills or resources with a new means of designing, fabricating, and personalising their equipment (Figure 6.48).

However, what about people who do not have advanced CAD skills? Novak also provided an algorithmic tool and detailed description to allow non-designers to customise surf and stand-up-paddle fins for 3D printing [27]. Features such as the fin system (e.g., Futures or FCS), depth, base length, cant, sweep/rake angle, and others can be modified using simple interactive tools and a 3D model that updates in real time (variations are illustrated in Figure 6.49a). Parametric relationships between different customisable features ensure that impossible geometries, such as those without any thickness that could not be 3D printed, cannot be accidentally created by users.

3D printed surf fins can also utilise novel materials, for example, carbon fibre-filled nylon using a Markforged 3D printer as in Figure 6.49b, which can be additionally reinforced with continuous carbon fibre to increase stiffness. Surfers may choose to experiment with such composites, as well as the infill percentages and patterns, to further tweak performance and create fins that could not be manufactured using conventional methods.

6.48
Comparison of a conventional mass-produced kitesurfing fin and 3D printed ABS version (left), 3D printed kitesurfing fin during early tests on the water (right) (Novak).

6.49
Example range of
surf fin shapes that
can be generated
algorithmically
using Grasshopper
and Rhino, with
users interacting
with a simple
control interface
that hides the
complex model
processes (a), 3D
printed carbon
fibre-filled nylon
surf fin, generated
algorithmically
(b) (Novak).

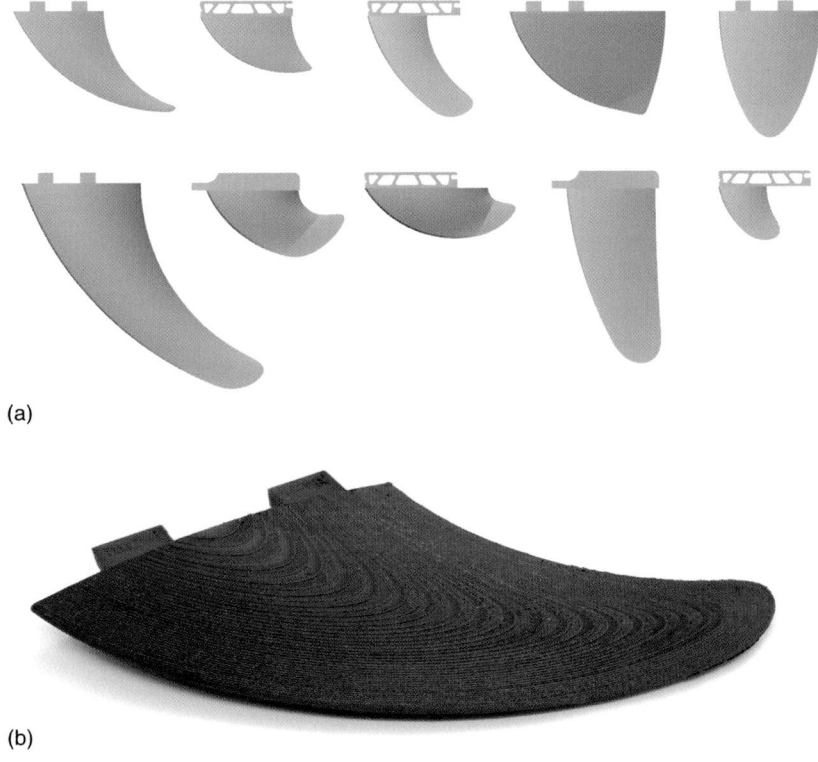

(a)

(b)

Heat exchangers

One of the most illustrative examples of using additive manufacturing for form-follows-function to create product innovation is the use of metal laser-based powder bed fusion for creating heat exchangers (Figure 6.50). However, this product also illustrates that as the technology enables complexity in fabrication, it requires increasing complexity in the development tools used to create the desired outcomes. With the growing maturity of metal additive manufacturing, new approaches to design software have been optimised for complex lattice structures. Conventional CAD software can be used but can have trouble in dealing with large arrays due to the required processing power. New software approaches are overcoming this challenge by handling features as equations (aka implicit modelling) rather than as solid bodies. Rather than develop a lattice from a series of operations to add and subtract material in CAD software, and then use patterning features to expand this model, the equation-driven approach makes it easier to design products with large, complex lattice structures, textures, and features without requiring high-end computing and graphics. Without this new breed of software, it would be difficult, if not impossible, to design these complex shapes. Also, a heat exchanger using gyroid structures would be impossible to produce without additive manufacturing.

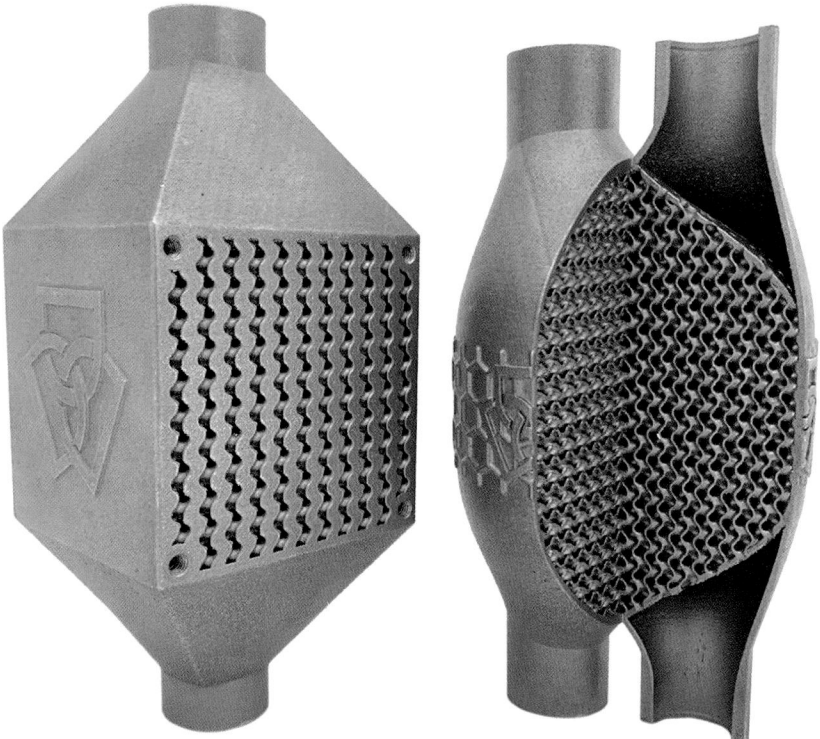

6.50
Cross section
through 3D printed
heat exchanger
(Diegel).

Heat exchangers and the gyroid revolution

A heat exchanger is a system used to transfer heat between two or more media. It allows heat from one substance, usually a liquid or gas, to pass to a second liquid or gas. The device is used to cool hot areas of a system or vice versa. The two media do not mix or come into direct contact with one another. Heat exchangers are commonly used in both cooling and heating processes for a wide range of industrial applications such as refrigerators, furnaces, air conditioning systems, transportation, oil refineries, commercial environments, and hospitals. Historically, "plate and frame" heat exchangers were made by forming plates into a labyrinth of channels. The plates were laminated to create a network of hot and cold channels to transfer heat from one media to the other. The other main conventional manufacturing option has been "shell and tube" heat exchangers. This type combines thermally conductive tubes and plates used to transfer heat from air to air, from water to water, or from air to water to steam.

A gyroid is part of a family of triply periodic minimal surfaces (TPMS), discovered by Alan Schoen in 1970. It separates space into two oppositely congruent labyrinths of passages that have no lines of reflectional symmetry. In the context of heat exchangers, gyroids have two distinct advantages. On the one hand, they offer a large surface area for heat dissipation. On the other, they

6.51
Illustrative example of a single gyroid cell (two versions on the left) and a network of connected gyroid cells (two versions on the right). The red (hot) zone is kept separate from the blue (cold) zone.

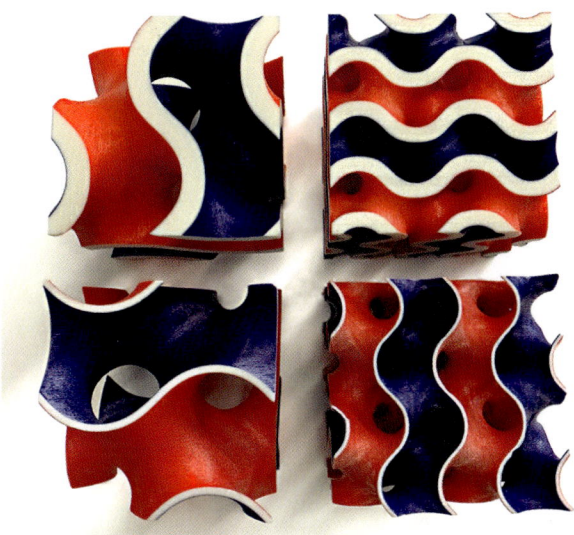

cause significant turbulence, increasing the Reynolds number, which improves heat dissipation. In the context of metal additive manufacturing, they have a distinct advantage of being self-supporting. Even at relatively large sizes, these structures can be printed without the need for support material. These advantages make gyroids a good candidate for creating more efficient heat exchangers (Figure 6.51).

Designing a gyroid heat exchanger

This project began as an exercise in understanding the capabilities of new equation-driven CAD design software, in this case nTopology. As an example, a compact heat exchanger was designed for a conformal water channel cooling system for a pellet extruder. As a first step, the physical structure of the heat exchanger was designed using Solidworks. New equation-driven CAD software products are capable of basic geometric features, but they are not as advanced as conventional CAD software for creating overall forms that define most products.

For this heat exchanger design, one side was modelled and then mirrored to produce the desired shape (Figure 6.52). The part was imported into the nTopology lattice design software. From this point, core bodies were created to make up the various parts of the heat exchanger. They included: (i) the cold and hot channel outer shells, (ii) the intersecting volume that would later become the gyroid space, and (iii) a few extra bodies such as the extra material required at the inlets and outlets for threads, logos, and more. The overall logic was to create all the subcomponents as separate units, called blocks, using nTopology. A range of Boolean operations (i.e., unite, intersect, and subtract) were used to combine the subcomponents.

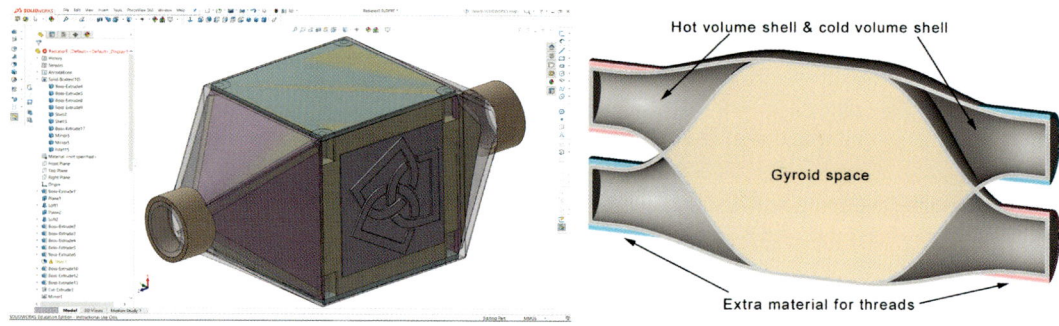

Hot volume shell & cold volume shell

Gyroid space

Extra material for threads

Creation of various heat exchanger bodies

After separating the individual subcomponent bodies, the gyroid-filled lattice space was created. A few heat transfer calculations were used to determine the surface area required to meet the desired conditions. This establishes the optimal gyroid cell size and wall thickness, which were then used to convert the lattice volume into a walled gyroid TPMS with the proper cell size and wall thickness.

Next, the gyroid structure must be "capped" to separate the hot and cold zones, and a solid TPMS volume is created (Figure 6.53). This represents the gyroid lattice that sits between the previously created gyroid walls. A few more Boolean intersection and union operations cap the cold channels on the hot side and vice versa.

Creating the hot and cold zone caps

Finally, a few further Boolean operations are used to combine all the subcomponents into the single ready-to-print heat exchanger. With some basic design-for-additive manufacturing techniques, it is possible to create extremely complex heat exchangers in which minimal support material is required when printing the part.

6.52
Basic CAD body for a radiator showing 10 separate bodies that form the radiator subcomponents (left) and a diagram of a heat exchanger (right).

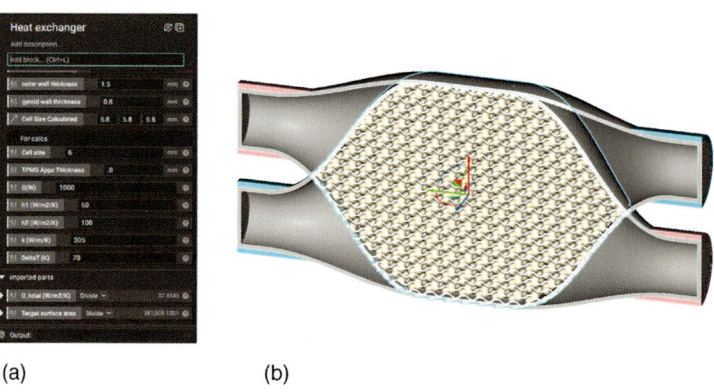

(a) (b)

6.53
Creation of a gyroid with a cell size and wall thickness to meet specified heat transfer characteristics (a) with a cross section of the TPMS (b).

6.54
Section view of completed heat exchanger, including hot and cold fluid zones (a), the printed part showing minimal support material requirements (b), printed heat exchanger section (c), and assembly for testing (d) (Diegel).

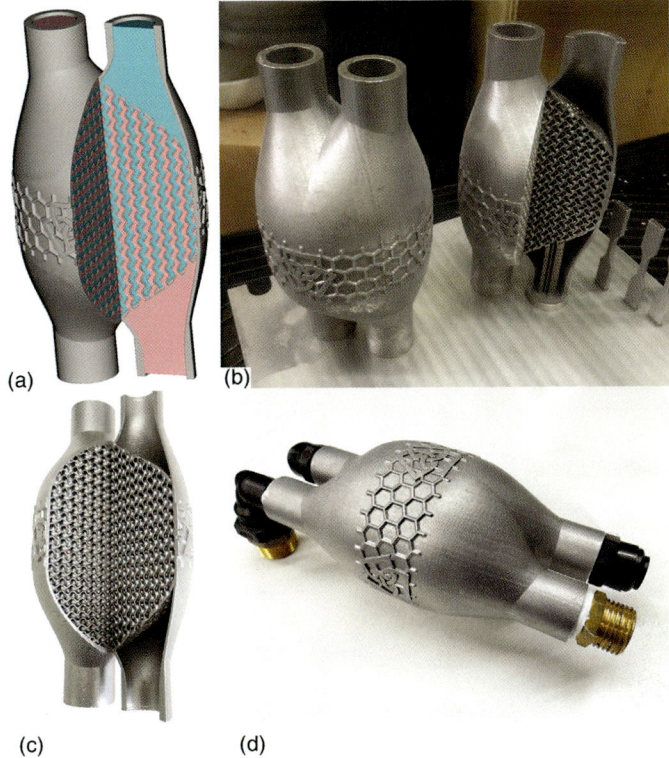

(a) (b)

(c) (d)

For this heat exchanger, the unsupported angle that is possible was pushed to the limit, with a small amount of support material used to ensure it printed correctly (Figure 6.54). Support material was required only in areas where removal was easy. Fortunately, the parts printed successfully on the first try. The initial test was running boiling water through the hot channel and ambient temperature water through the cold channel. A temperature differential of about 65°C (149°F) was achieved. Detailed simulations and more scientific measurements are expected in the future.

Creating the next generation of heat exchangers

With this new equation-driven design software, coupled with the proper workflow, one can almost instantly create a new heat exchanger with different characteristics. This makes it ideal for researching more efficient forms of heat exchange. Another example shows the transition from a fluid-to-fluid heat exchanger to an air-to-fluid heat exchanger (a radiator) using a similar workflow.

In summary, once a workflow has been created, it is possible to quickly produce gyroid-based heat exchangers of different sizes and efficiencies (Figure 6.55). With a few modifications, the workflow can be adapted to create other

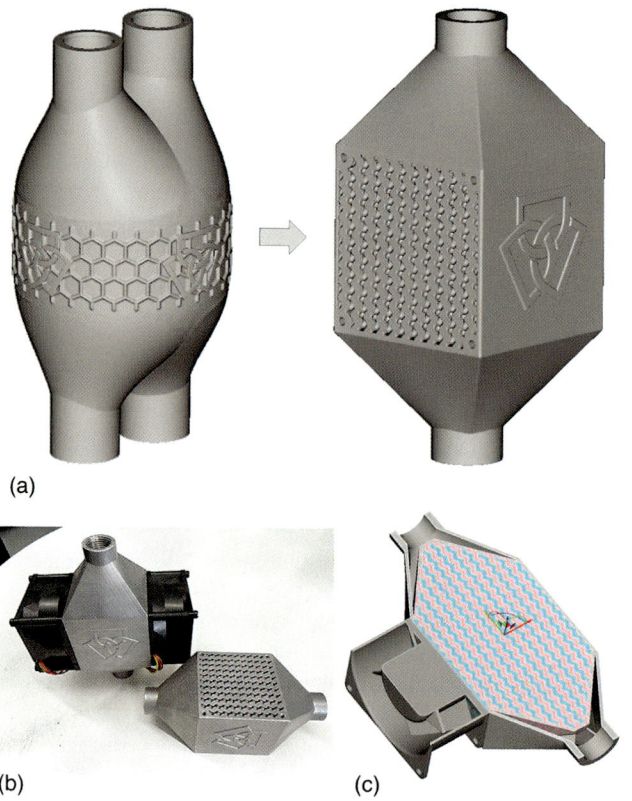

6.55
Using a related workflow to go from a heat exchanger to a radiator (a). A printed radiator is made using the same heat exchanger workflow (b) and cross-sectional view (c).

(a)

(b)   (c)

devices, such as radiators. This project demonstrated the power of gyroids and how they can be used as self-supporting structures to transfer heat from one substance to another. With a large surface area, these designs are excellent candidates for heat exchangers and for creating lightweight self-supporting structures with metal additive manufacturing.

REFERENCES

1. Novak JI, Novak AR. Is Additive Manufacturing Improving Performance in Sports? A Systematic Review. *Proceedings of the Institution of Mechanical Engineers, Part P: Journal of Sports Engineering and Technology* 2020;235(3):163–175.
2. Renishaw. First Metal 3D Printed Bicycle Frame Manufactured by Renishaw for Empire Cycles: Renishaw; 2014 [updated 04/02/2014]. Available from: http://www.renishaw.com/en/first-metal-3d-printed-bicycle-frame-manufactured-by-renishaw-for-empire-cycles--24154.
3. Novak JI. *A Study of Bicycle Frame Customisation Through the Use of Additive Manufacturing Technology. RAPID 2015.* Long Beach, CA: SME; 2015.
4. Berg E. Designing for Performance and Protection with Digital Manufacturing. In: DelVecchio SM, editor. *Women in 3D Printing: From Bones to Bridges and Everything in Between.* Cham: Springer International Publishing; 2021. pp. 19–31.

5. Novak J, Burton D, Crouch T. Aerodynamic Test Results of Bicycle Helmets in Different Configurations: Towards a Responsive Design. *Proceedings of the Institution of Mechanical Engineers, Part P: Journal of Sports Engineering and Technology* 2019;233(2):268–276.

6. Kantaros A, Diegel O. 3D Printing Technology in Musical Instrument Research: Reviewing the Potential. *Rapid Prototyping Journal* 2018;24(9):1511–1523.

7. Pye, D. *The Nature and Art of Workshop*. London: Bloomsbury; 2006.

8. Dean LT, Atkinson P, Unver E. Evolving Individualised Consumer Products. *6th European Academy of Design Conference*; Bremen, Germany. 2005.

9. Novak JI, O'Neill J. A Design for Additive Manufacturing Case Study: Fingerprint Stool on a BigRep ONE. *Rapid Prototyping Journal* 2019;25(6):1069–1079.

10. Dean L, Loy J. Generative Product Design Futures. *The Design Journal* 2020;23(3):331–349.

11. Giaccardi E. Metadesign as an Emergent Design Culture. *Leonardo* 2005;38(4):342–350.

12. Fuad-Luke A. *Design Activism: Beautiful Strangeness for a Sustainable World*. London: Taylor and Francis; 2009.

13. Leach N, Farahi B, editors. *3D-Printed Body Architecture, Architectural Design*, 06, V87; 2017.

14. Barrios-Muriel J, Romero-Sánchez F, Alonso-Sánchez FJ, Rodríguez Salgado D. Advances in Orthotic and Prosthetic Manufacturing: A Technology Review. *Materials* 2020;13(2).

15. Jacobs S, Schull J, White P, Lehrer R, Vishwakarma A, Bertucci A, editors. e-NABLING Education: Curricula and Models for Teaching Students to Print Hands. *2016 IEEE Frontiers in Education Conference (FIE)*; 12–15 Oct. 2016; Erie, PA, USA: IEEE.

16. Sorrel C. Bespoke Designs Makes Beautiful Custom Prosthetic Legs. *Wired*; 2010. Available from: https://www.wired.com/2010/12/bespoke-designs-makes-beautiful-custom-prosthetic-legs/.

17. Asfia A, Deepak B, Novak JI, Rolfe B, Kron T. Multi-Jet Fusion for Additive Manufacturing of Radiotherapy Immobilization Devices: Effects of Color, Thickness, and Orientation on Surface Dose and Tensile Strength. *Journal of Applied Clinical Medical Physics* 2022;23(4):e13548.

18. Puls N, Carluccio D, Batstone MD, Novak JI. The Rise of Additive Manufacturing for Ocular and Orbital Prostheses: A Systematic Literature Review. *Annals of 3D Printed Medicine* 2021;4.

19. Pullin G. *Design Meets Disability*. Cambridge, MA: MIT Press; 2009.

20. Dickson CD. *From Stigma to Silver Linings: Improving the Experiences of Long-Term Tracheostomy Users Through Product Design*. Auckland: Auckland University of Technology; 2017.

21. Maclachlan LR, Alexander H, Forrestal D, Novak JI, Redmond M. Properties and Characteristics of 3-Dimensional Printed Head Models Used in Simulation of Neurosurgical Procedures: A Scoping Review. *World Neurosurgery* 2021;156:133–146.

22. Asfia A, Novak JI, Mohammed MI, Rolfe B, Kron T. A Review of 3D Printed Patient Specific Immobilisation Devices in Radiotherapy. *Physics and Imaging in Radiation Oncology* 2020;13:30–35.

23. Asfia A, Novak JI, Rolfe B, Kron T. Development of a Patient-Specific Immobilisation Facemask for Radiation Therapy Using Additive Manufacturing, Pressure Sensors and Topology Optimisation. *Rapid Prototyping Journal* 2021;28(5):945–952.

24. Desselle MR, Kirrane M, Chao IT, Coles Black J, Woodruff MA, Chuen J, et al. Evaluating the Safety and Effectiveness of Novel Personal Protective Equipment during the COVID-19 Pandemic. *Medical Journal of Australia* 2021;214(11):496–499.e1.
25. Yuan P, Menges A, Leach N. *Digital Fabrication*. London: University Press; 2007.
26. Tibbits S, editor. *Active Matter*, Cambridge, MA: MIT Press; 2017.
27. Novak JI. A Parametric Method to Customize Surfboard and Stand-Up Paddle Board Fins for Additive Manufacturing. *Computer-Aided Design and Applications* 2021;18(2):297–308.

7 DfAM

Design guidelines for product designers

The intent of this chapter is to be a resource for designers working with the most common additive manufacturing technologies for product applications, aligned with the three strategies outlined throughout this book. It includes quick-reference graphics and tables for key features and specifications relevant to leading technologies, materials, and applications. The constraints relevant for the families of 3D printing technologies are described, as well as practical design guidelines for some of the key features of product design, with comparisons between additive technologies clarifying their opportunities and challenges.

Pre- and post-processing considerations also influence design decisions, and these are outlined to provide further design guidelines for specific additive technologies. This covers the end-to-end workflow from product design through part optimisation, print preparation, and final part finishing. However, it is not an exhaustive technical guide and does not detail all of the main families of additive technologies, which has been covered in numerous books (e.g., [1]) and online resources. Instead, this chapter bridges the gap between the opportunities of additive technologies and the reality of working with them hands-on within a professional manufacturing environment.

The implication of the technology overview section is that professional designers and engineers can build up their knowledge of the main additive technologies by beginning with the least restrictive polymer powder bed fusion (PBF) processes and working through material extrusion to metal PBF. While material extrusion is the most accessible and affordable technology utilised by many designers, it has an increased number of considerations and restrictions compared with polymer PBF. This makes it more challenging to design for, and more challenging to integrate into end-use applications aligned with the strategies of this book.

Importantly, this chapter is a guide to inform thinking and highlight some of the most important design considerations and their associated business ramifications. Costs, dimensions, tolerances, and outcomes are highly dependent on the specific additive hardware, materials, and process parameters being used. With the rapid pace of technology development, they are frequently changing

DOI: 10.4324/9781003122203-8

and, thus, it is strongly recommended that designers experiment with each technology to understand the limits of what is possible, and that they question "rules" specified in manuals or by manufacturers.

OVERVIEW OF SEVEN FAMILIES OF ADDITIVE MANUFACTURING TECHNOLOGY

Before the chapter delves into the key polymer powder bed fusion, material extrusion, and metal powder bed fusion processes, the following is a summary of the seven families of additive technologies described by ISO/ASTM 52900 [2].

Vat photopolymerisation

In 1986, Chuck Hall was granted a patent for Stereolithography (SLA), a process that created 3D objects by successively curing layers of liquid resin using ultraviolet (UV) laser light. Hall founded the company 3D Systems, which has been a leader in the industry since that time.

Other processes within the family of vat photopolymerisation are Digital Light Processing (DLP), which utilises a UV projector light source rather than a laser one to cure an entire layer at a time, and Continuous Liquid Interface Production (CLIP). The CLIP system is a particularly interesting process; see Carbon3D (www.carbon3d.com/3d-printer-models-carbon/our-technology/). In all three, the 3D model is sliced and cured in layers. Resin printers can print from the bottom plate upwards, or the build can be printed hanging off a top build plate. Vat photopolymerisation is suitable for high-resolution parts due to the very fine layer thickness. Many of the uses exploit the optical clarity possible in the parts, as well as an increasing variety of biocompatible or flexible material options.

Vat photopolymerisation processes print in one material only and, because they are built on a platform, require support scaffolding for overhangs, like those needed in single-filament material extrusion. "Good design" with this technology, therefore, minimises or preferably avoids support scaffolding in the print except to hold the part to the build platform. Whilst most SLA, DLP, and CLIP systems are desktop, there are also massive SLA machines with dimensions measured in metres.

Powder bed fusion

The term "powder bed fusion" covers several technologies that all involve powder, spread across the build plate in a thin layer, being fused in some way. This can be either by laser (selective laser sintering or selective laser melting), electron beam (melting), or infrared heat (as in multi jet fusion). These technologies are suitable for polymers and/or metals. One of the key design considerations is that the prints are supported by the surrounding vat of powder once each layer is printed. This means that complex geometries, overhangs, and assemblies can be printed as a single part, and support structures are required only for metal parts to aid with heat dissipation and minimise warping.

Multi Jet Fusion (MJF) is a relatively recent technology, developed by Hewlett Packard (HP). This uses "inks," similar to an inkjet printer, to mark out the external boundaries and interior structures of a print, and a heating element to solidify whatever areas have had ink applied to them in the model [3].

Binder jetting

Binder jetting is one of the earliest forms of 3D printing. It involves powdered material, such as plaster, spread across a build plate, with inkjet-printed droplets of adhesive binder solidifying the form on each layer. Binder jetting can be a single-colour or a full-colour print process.

Material jetting

This is a 3D printing process that features multiple print heads that can blend and jet tiny droplets of UV-curable resin, allowing for full-colour and multi-material parts. The resin is cured by a UV light mounted to the print head, solidifying the material as soon as it is deposited so that additional layers can continue to be printed on top. Some of the materials used in this process may be dedicated support material, which may be soluble or melted when in a wax form, simplifying post-processing.

Sheet lamination

Sheet lamination is a less common technology in the context of additive manufacturing, but it technically sits within the definition. In this process, sheets of material (e.g., paper) feed into the machine, with each layer cut and glued on top of the previous one.

Material extrusion

Fused filament fabrication (FFF), known more commonly as fused deposition modelling (FDM), a trade name registered by Stratasys in its desktop form, is the most widespread 3D printing technology used for prototyping in the community, as it is the least expensive. It essentially works like a hot glue gun, extruding a continuous filament to build the model on a build plate. The filament is heated to bond each layer. The simplest form of the technology features a single extruder allowing it to print with a single filament at a time. "Good design" using this technology reduces or avoids the need for support structures, as these are made using the same filament and, therefore, can be difficult to remove in complex geometries. However, dual- or multi-filament FFF can use a soluble support structure that can simplify the post-processing. It can also allow for multi-colour and multi-material printing. There are also different versions of the technology, including those designed for vertical printing or those that print on treadmill-style rotating beds that significantly extend the build volume in one plane. The extension of these is 3D printing using a robot arm. In industry, high-end machines are predominantly dual-filament machines that create precision outputs suitable for commercial applications, with high-temperature materials. The cost of support material must also be accounted for in FFF industrial applications.

Directed energy deposition

These technologies (LENS, DMD, LMD) have more in common with welding than some other versions of additive manufacturing, as powder or wire is fed into a melt pool of metal. The process tends to be less accurate than other additive manufacturing technologies and, therefore, is used for different applications. CNC machining or other subtractive processes are commonly used to clean up rough surfaces from these processes.

DIVING DEEPER INTO THE THREE MOST COMMON ADDITIVE TECHNOLOGIES FOR PRODUCT DESIGNERS

Polymer powder bed fusion

Polymer powder bed fusion (PBF) processes include selective laser sintering (SLS) and high-speed sintering (HSS) MJF. The benefit of these processes is that parts remain suspended within the unfused powder as they print, requiring no additional support materials or attachment to the build platform common to most other additive technologies. This provides significant design and manufacturing freedom and is one of the key reasons polymer PBF makes the ideal starting point for commercial design applications, particularly where upskilling of employees is required. The second advantage of polymer PBF technologies is that parts can perform visually and functionally like similar products of the same material produced using traditional manufacturing – for example, a good surface finish, and mechanical properties that are more homogenous than those produced using other layer-by-layer processes, such as FFF. However, parts can still exhibit anisotropic qualities if not designed properly [4]. This means that there are differences in mechanical performance in the different XYZ orientations. While the surface finish of PBF parts is not comparable to injection moulding, exhibiting a fine sandy texture, the final parts are acceptable for many end-use applications, and post-processing can improve surface quality.

As a more established process, SLS accommodates a broader range of materials than MJF, as shown in Table 7.1.

Another key difference between parts produced by SLS and MJF is the colour of the final part. SLS is currently limited to producing parts in the same colour as the raw material input, simply fusing the material with a laser. For the most common polyamide materials, this is normally white. While SLS parts can be dyed or painted, additional equipment and manual labour is required after

Table 7.1 Commonly available materials for SLS and MJF

SLS	MJF
Aluminium-filled nylon (Alumide)	Polyamide 11 & 12 (aka Nylon)
Carbon fibre-filled nylon	Thermoplastic urethane (TPU)
Glass-filled nylon	
Polyamide 11 & 12 (aka Nylon)	
Polystyrene	
Thermoplastic urethane (TPU)	

printing. On the other hand, despite MJF also using polyamide material, it deposits a black fusing agent to absorb heat and cause powder fusion. As a result, all parts come out of the machine in a dark grey colour. While post-processing technologies are emerging to provide colour options, the most common post-process is to dye the parts black, or simply leave them in their raw dark-grey state.

HP explored the potential of creating a colour range through MJF machines that provided additional detailing agents around the perimeter of parts during printing. This resulted in a white surface finish that was 0.5 mm thick, while the internal fusing agent remained dark grey. This white finish could also absorb CMYK inks during printing, allowing full-colour parts to be produced within a single printing process, as shown in Figure 7.1. However, most of the interior of the printed part was dark grey. This technology was interesting from a product designer's point of view, because of the potential to add detailing using the colour facility. However, because the colour was a surface finish only, it was problematic for end-use parts because of the potential for the colour to be worn away. Additional surface finishes could be added as sealants, but at the time of writing this book, there was some uncertainty about the ongoing availability of the colour facility for MJF, which would be a loss from a product designer's perspective.

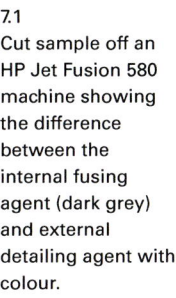

7.1
Cut sample off an HP Jet Fusion 580 machine showing the difference between the internal fusing agent (dark grey) and external detailing agent with colour.

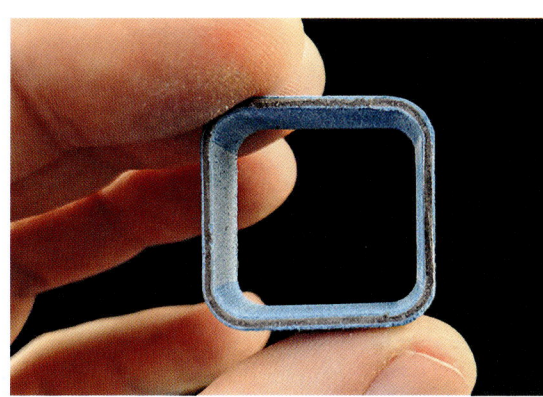

Table 7.2 Opportunities and challenges of PBF in relation to the three strategies of this book

	Opportunities	Challenges
Strategy 1	Prototypes can look and feel similar to mass-produced parts (although some post-processing may be required). Functional materials are useful for bridge manufacturing and batch production applications.	High hardware cost means it is unlikely a business would buy one of these technologies for prototyping or low-volume production. Use of service bureaus is likely. Large flat surfaces are susceptible to warping; consider stiffening them with ribs where the design allows.
Strategy 2	Complex geometries can be fabricated without support material. Fully assembled parts can be manufactured. Suitable for mass-customisation systems supported by software to automate nesting of many different parts within a build volume. Machines can run 24/7 with minimal supervision.	Powder bed processes are limited to single materials, so different materials still require assembly. Manual processes are still required in pre- and post-processing which must be factored into costs. Build volumes are relatively small compared to other additive technologies, limiting product size. Limited variety of polymers available, with polyamide and TPU the most common. Parts are slightly porous and require post-processing and optimised process parameters in order to minimise absorptivity.
Strategy 3	Proven technologies to work across global 3D printing networks, making new business models and supply chains possible. Print times are relatively fast (~8 hours per build), making print-on-demand applications e.g., digital inventory, possible.	Waste powders can be difficult to recycle, especially depending on location and whether suppliers provide a return program.

Due to the maturity of polymer PBF technology over several decades, many commercial opportunities have been realised. These are summarised in Table 7.2 and have been aligned with Strategies 1, 2, and 3 discussed in Chapters 3, 4, and 5. However, like all additive technologies, there are limitations that businesses and manufacturers must consider, including materials, surface quality, build volume, and hardware cost.

Material extrusion

As discussed in Chapter 1, material extrusion is the most ubiquitous method of 3D printing, with the most common form of the technology known as FFF or FDM. FFF technology involves heating a polymer filament and extruding it through a nozzle, which is computer-controlled to trace the profile of a part.

Despite being the most common 3D printing technology utilised by most designers and engineers, those who have physically operated a FFF 3D printer will know that the technology comes with numerous challenges. This includes the inherent need for support material, challenges with material warping, anisotropic parts, visible layer steps, and relatively slow production speeds. For these reasons, material extrusion is a more challenging technology to design for compared to polymer powder bed fusion. Within the framework of the three strategies discussed in Chapters 3, 4, and 5, material extrusion may not be the best starting point for individuals and businesses within a commercial context.

Material extrusion is a broad category which includes numerous variations on the FFF process, including the ability to print with metals in a green state prior to a secondary process of sintering in a furnace, known as "bound metal deposition." A more detailed breakdown of materials for FFF and metal extrusion is available in Table 7.3.

Non-heated methods of 3D printing, with materials like clay, concrete, foods, biomaterials, and just about any other material that can be forced through a nozzle, are also possible [5]. This also includes the use of robot arms and other mobile robots. While many of these may be of interest to designers, the focus of this chapter will be on FFF, with polymers and composites the most common materials product designers use with the widest range of applications. These are summarised in Table 7.4.

Even within FFF, hardware and material costs, as well as layer thickness, accuracy, and other parameters, vary wildly between brands and specific machines. Generally, you get what you pay for. For example, a US$350/300 pounds sterling desktop machine will generally be unable to produce parts to

Table 7.3 Commonly available materials for FFF and metal extrusion

Polymers for FFF	Metals for bound metal deposition
Acrylonitrile butadiene styrene (ABS)	Copper
Acrylonitrile styrene acrylate (ASA)	Inconel
Polyamide 11 & 12 (aka Nylon)	Steel
Polycarbonate (PC)	Stainless steel
Polyether ether ketone (PEEK)	Tool steel
Polyetherimide (PEI)	
Polyethylene terephthalate (PET)	
Polyethylene terephthalate glycol (PETG)	
Polylactic acid (PLA)	
Polypropylene (PP)	
Polyvinyl alcohol (PVA, for soluble supports)	
Thermoplastic urethane (TPU)	

Table 7.4 Opportunities and challenges of FFF in relation to the three strategies of this book

	Opportunities	Challenges
Strategy 1	Accessibility and affordability of desktop FFF machines makes it suitable for any design, engineering, or manufacturing space. This is perfect for quick prototypes and iterative design.	Rough surface finish of prototypes will not replicate details of traditional manufacturing.
	Large build sizes can produce large jigs and fixtures, as well as moulds, in a single print.	Support material is typically a manual process to remove, costing time and money.
Strategy 2	Large build sizes are suitable for large objects like furniture, sculptures, or other items as large as buildings.	Extrusion technologies are comparatively slow compared to other additive processes.
	Multi-colour and multi-material parts are possible.	Rough surface finish makes it unsuitable for many end-use parts. Post-processing adds time and cost.
	Large variety of materials available to meet a variety of end-use part requirements.	Support material is typically a manual process to remove, costing time and money and negatively affecting surface finish.
	Multiple 3D printers can be networked together to operate as a build farm, enabling mass customisation and improving production volumes.	The anisotropic properties of parts produced through extrusion can make them unsuitable for many end-use applications.
Strategy 3	With increasing availability of FFF technology within communities, new business opportunities are possible where designs are sent to local providers to be manufactured close to customers.	Quality control with FFF 3D printers is particularly challenging, and the ideal of digitally sending a file to different printers around the world and expecting them to perform the same will limit new business models.
	Low-cost filament extruders and new FFF 3D printers that utilise waste plastics as feedstock enable new closed loop production.	As more people have access to these 3D printers, more plastic waste may be produced with people printing trinkets and throwaway items downloaded from online.
	More sustainable materials, or materials made from waste products, present new business opportunities.	Regulations and intellectual property have been slow to adapt to the capacity of desktop 3D printing, and in many industries may restrict innovation.
	As seen through the COVID-19 pandemic, the ubiquity of FFF 3D printers can be used by the community to supply emergency products such as personal protective equipment when traditional supply chains break down.	

the same standard as a US$3500/3000 pounds sterling machine, even if they can print with the same layer thickness or other features. However, spending more money is also not a guarantee of quality. Navigating the material extrusion category can be challenging for businesses due to the sheer volume of hardware options on the market, particularly at the lower price point, i.e., <US$3500/3000 pounds sterling or even the <US$700/600 pounds sterling category. Strategies to help in researching a machine prior to purchase include:

- Connect with colleagues in other businesses who have purchased and used extrusion 3D printers to establish who they purchased from, how the after-sales support was, whether they would make the same purchase again, regularity of 3D printer use, and the new opportunities discovered.
- Reach out to local universities or makerspaces with 3D printing capability as people in these spaces often have access to a broad range of 3D printers and are eager to share both positive and negative feedback.
- Reading online reviews and forums, including through social media with many groups formed around particular machines or brands that can be valuable for troubleshooting.
- Search for award-winning 3D printers; for example, websites like 3D Printing Industry, Make Magazine, All3DP, and 3D Hubs, which annually review and award the best 3D printers of the year in several categories. These are typically for consumer machines rather than high-end commercial machines.

Metal powder bed fusion

Metal powder bed fusion (PBF) falls within the same ISO/ASTM 52900 standards for additive manufacturing as polymer PBF. However, due to the increased complexities of design, pre-processing, and post-processing with metal PBF, it is recommended they be considered as separate technologies with distinct applications, as identified in Chapter 1. These challenges also make it difficult to learn as a first additive technology, and polymer PBF should be used as a starting point to learn fundamental principles before transitioning to metal PBF. In fact, metal PBF has a lot in common with FFF due to the need for support structures anchoring parts to the build plate.

Other than printing speed and materials (as listed in Table 7.5), the main differences between DMLS/SLM parts and EBM parts are that DMLS/SLM parts tend to have higher accuracy due to the precision of the laser, and a better surface finish due to the thinner layers. These are possible because of the use of finer metal powders compared to EBM powder particles.

Unlike polymer PBF, metal PBF parts require supports connected to the build plate. This performs two functions. First, they anchor the parts directly to the build plate to minimise the risks of warping or movement caused by the rapid heating and cooling of the fusing processes. Second, the supports draw heat away from the parts to further minimise heat build-up and warping. This means

Table 7.5 Commonly available materials for DMLS/SLM and EBM

DMLS/SLM	EBM
Aluminium	Cobalt chromium
Cobalt chronium	Inconel
Copper	Nickel alloy
Inconel	Titanium
Precious metals (e.g., gold, palladium, platinum, silver)	
Stainless steel	
Titanium	
Tool steel	
Tungsten	

that designers cannot entirely avoid the use of support material through clever design, although good design can help simplify the removal of support material. The need for supports also reduces the ability to stack parts through the entire build volume, sharing some similarities with material extrusion or vat photopolymerisation.

The main opportunities and limitations of metal PBF are summarised in Table 7.6. For businesses investing in this technology, these must be considered alongside the significant safety requirements for the technology. These include the safe storage and handling of fine metal powders (which can be explosive) and bottled gases, upgrading of safety equipment such as fire extinguishers, and ventilation. Further, depending on local regulations, the potential need for blast-proof screens/walls or other structural upgrades is a key consideration. It is common for the installation of a metal PBF machine to require moderate to significant renovation of a factory or fabrication space, increasing the initial investment in this technology.

DESIGN GUIDELINES FOR ADDITIVE MANUFACTURING

The significant breadth of hardware and materials available for additive manufacturing makes it challenging to define specific design rules. As such, this section discusses some of the overarching guidelines that must be considered during the design of parts for additive manufacturing, comparing the different core technologies of the previous section, and explaining the concepts behind limitations. While some dimensions are provided, these can always be challenged and are highly dependent on hardware, material, and process parameters. It is important to be wary of strict rules provided by manufacturers, service bureaus, or educators, as these are often based on operational efficiencies and commercial economies, with "safe" minimums and dimensions provided that ensure successful prints. This should not limit design. Experimentation and physical testing become key to realising the full potential of additive manufacturing.

Table 7.6 Opportunities and challenges of metal PBF in relation to the three strategies of this book

	Opportunities	Challenges
Strategy 1	Functional prototypes that perform similar to cast/machined metal parts. Suitable for metal tooling with improved performance features, e.g., conformal cooling. Durable and high-strength jigs and fixtures.	Expensive technology which may be better outsourced for most Strategy 1 applications rather than purchased for in-house use. Costs for simple jigs or moulds may be significantly higher than machined parts. Strategy 1 is normally a first step into the use of additive manufacturing, and metal PBF may require much greater investment in the technology and/or upskilling workers than other processes.
Strategy 2	Functional end-use parts that can share almost identical characteristics to machined or cast metal parts. Good range of metals for many applications with new alloys being developed specifically for additive manufacturing.	Post-processing is required to achieve a similar surface finish to parts made through other metal technologies. As a minimum, manual finishing is required to remove support material. Challenging for designers and engineers to get hands-on experience with this technology compared to many others. Part sizes are limited by often small build volumes compared to other technologies.
Strategy 3	Shipping metal parts globally is expensive due to weight – being able to send digital files for manufacture closer to where parts are needed is changing supply chains. Service bureaus specialising in metal PBF can play a role in helping small-to-medium enterprises (SMEs) utilise the technology without needing to buy a machine. Unused metal powders can be collected within the machine for re-use, minimising waste.	Waste is unavoidable through the need for support material. With many opportunities come many barriers to adopting parts fabricated through metal PBF, including regulators in tightly regulated industries like medicine and aviation.

DESIGNER'S COMMENT

Design for Additive Manufacturing is not a choice. It is an absolute necessity if you want to add true value to your products.

Build volume

As with any manufacturing technology, product and part dimensions are constrained by the hardware and processes of manufacturing. Within additive manufacturing, build volumes vary significantly between technologies, as well as within each technology type. Therefore, designers need to understand the end manufacturing build volume (or envelope) to inform design decisions as early as concept development. This may be less critical for some Strategy 1 scenarios, such as prototypes which can be segmented into pieces to be assembled later. However, they will be more important for Strategy 2, where focus is on end-use parts or competitive markets, and which requires minimising manual assembly costs.

Table 7.7 provides an overview of build volumes for the key additive technologies in this chapter. As a rule, the larger the build volume of a machine, the more expensive it is to purchase. It is worth noting that build volumes are not always cubic, and it is common for one dimension to be greater than the others, perhaps in a highly exaggerated way, providing a rectangular or even cylindrical build volume. An extreme example of this in the desktop FFF category are 3D printers featuring treadmill-style belt platforms, as opposed to a fixed plate, which, theoretically, provides an almost limitless dimension in one horizontal orientation.

PBF technologies offer some of the smallest build volumes for any additive manufacturing technology, whereas material extrusion provides the largest build volumes suitable for vehicles, buildings, and other large structures. The extra-large FFF dimensions shown in Table 7.7 are based on hardware that product designers may utilise. However, material extrusion technologies can have build

Table 7.7 Overview of build volumes for key additive technologies

Size	Polymer PBF (mm)	Desktop FFF (mm)	Commercial FFF (mm)	Metal PBF (mm)
Small	100 × 100 × 100	100 × 100 × 100	254 × 254 × 254	100 × 100 × 100
Medium	300 × 300 × 350	250 × 210 × 210	406 × 355 × 406	300 × 200 × 200
Large	550 × 550 × 750	500 × 500 × 500	1005 × 1005 × 1005	800 × 400 × 500
Extra Large	N/A	N/A	3556 × 1651 × 914 up to 6096 × 2286 × 1829	N/A

volumes measured in tens of metres for building and construction applications. For large and extra-large FFF build volumes, nozzle diameters typically increase from the common 0.4 mm (±0.2 mm) of smaller machines to 1 mm or greater. This correspondingly increases the thickness of layers and minimum feature size, and while it may be possible to use smaller nozzles on these large machines, the time to print a large object significantly increases with smaller nozzles and layer thicknesses [6].

A simple tool to help reinforce a build volume limitation is to model a box matching the build volume within CAD and use this as a wireframe reference to design parts within. Several tricks to extend the size of parts possible within a build volume include:

- Making use of the diagonal space within a build volume rather than being restricted to the maximum XYZ dimensions. This works better for PBF than material extrusion, where significant supports are required.
- Creating flexible forms that allow a single design to be digitally folded or hinged within a build volume. This works particularly well for fabric-like structures and means large surfaces can be 3D printed in a relatively small build volume.
- Dividing large parts into smaller parts that then get glued together. Most STL file manipulation software allows for this and enables the use of auto-matically created lap joints or male/female pins and holes.

Wall thickness

The minimum wall thickness is a highly variable property and will be influenced by the additive hardware, material, and process parameters, as well as the functional requirements of a part. Theoretically, the minimum thickness for a wall is determined by the fusing mechanism. This ranges from the diameter of the laser in SLS or SLM, the size of a voxel in MJF, or the diameter of the nozzle in material extrusion. In most scenarios, this value will not provide a functional wall for a product. However, it may provide limits for small details like ribs or text, and careful fine-tuning is necessary to utilise minimum dimensions.

For polymer PBF technologies, walls that are a minimum of 0.7 mm are recommended, with anything over 1 mm being relatively safe and reliable. Nevertheless, walls 1–5 mm thick are also common, although thicker sections up to 20 mm are possible if excess material is removed, and the part is oriented to minimise warping during printing. For metal PBF, the minimum wall thickness is smaller, usually at ≥0.3 mm. However, a wall thickness of 1 mm or above is also a safe minimum to work with for most wall sections.

Material extrusion is more difficult to prescribe, with wall thickness being a complex relationship between the nozzle diameter, layer height, printing speed, and the flow rate of material going through the nozzle. A simple rule to remember for the minimum wall thickness is to add 20 percent to the nozzle diameter.

For example, a standard 0.4 mm nozzle will have a minimum wall thickness of 0.48 mm (or 0.5 mm to keep dimensions simple). The thickest a nozzle can extrude in a single pass is 2x the nozzle diameter; so, 0.8 mm for a 0.4 mm nozzle. Of course, anything over the minimum dimension will print and can be accommodated by a combination of perimeter and infill extrusions, with no real limit to maximum wall thickness for FFF.

While designing to the minimum wall thickness is possible with material extrusion and will be used for small details, it does have limited applications due to the limited surface area adhering each layer to the next, resulting in low strength. These applications include quick hollow prototypes (often known as printing in "vase" or "eggshell" mode) and more artistic products like vases and sculptures, popular when printing in clay. The resolution of a STL or other mesh file for 3D printing will also affect minimum wall thickness.

Joining details

In most cases where a functional part needs to be larger than the available build volume, a different additive technology capable of large 3D printing will be more suitable. However, for some Strategy 1 applications, joining several 3D printed sections together may be appropriate where strength may not be as important (prototypes), or where additional details can be accommodated to reinforce joint areas without concern for the effect on aesthetics (jigs and fixtures). It is recommended that joinery details be added to aid in fixing and alignment, rather than simply slicing parts down to size with flat faces. Joints common in wood-working can be useful to include, and some software specific to 3D printing now accommodates for these to be used when separating a part into pieces for printing. Fasteners and other mechanical joining methods may also be suitable, including the use of threaded inserts commonly used in moulded plastic products. These can be inserted into holes with heat (e.g., soldering iron) as part of post-processing.

Lattice structures

For products developed as part of Strategy 2, more complex lightweight forms are likely, including the use of lattice structures. Lattice structures are repeating (tessellating) patterns of a frame-like geometry that can significantly reduce the weight of a part compared to solid material. This geometry can be defined within a unit cell and consists of nodes and beams (also called struts) [7]. The node is a joint that may fall within a unit cell, or at the perimeter where it will connect to other nodes as the unit cell repeats in 3D space. A beam connects two nodes and is normally cylindrical in shape. These could be designed manually in traditional CAD software and patterned throughout an area (as in Figure 7.2), or, increasingly, could rely on dedicated lattice tools to automate the generation of lattice structures throughout a part, or in specific zones.

Functional lattice structures are best suited to additive technologies with the ability to print small details with good adhesion between layers

7.2
3D FFF printed
lattice structures.

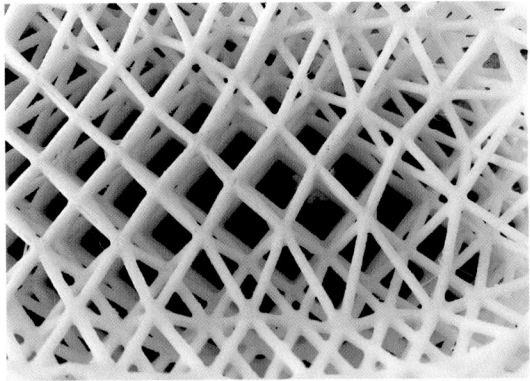

approaching isotopy. For these reasons, material extrusion is not ideal for functional lattice structures, as their thin struts are easily broken due to the weakness between layers and frequent movement of the nozzle from one strut to the next. Beyond the geometry of the lattice, which will affect mechanical performance, the other design criteria affecting the functional properties of a lattice are the beam diameter and the size of the unit cells. The beam diameter should follow the same guidelines as the wall thickness and, through advanced finite element analysis, may vary throughout a part, being thicker in areas under high load and thinner where loads are lower and part weight can be reduced. Such advanced lattice design may require additional software beyond standard CAD packages.

The size of unit cells will affect the overall density of a part, as well as the functional performance. Again, simulation software can help inform and optimise the unit cell size, although consideration must be given to the gap (or negative space) around the struts. A minimum gap of 5 mm is recommended between struts within a unit cell to allow loose powder or other material to be easily removed. Depending on the quantity of unit cells through a thick section of a part, the minimum gap may need to be larger to simplify the removal of powder deep within an area of lattice, where it can be difficult for blasting media and manual tools to reach. For metal PBF, consideration also needs to be given to the location of supports and the ability to remove them.

Assembled/moving parts

One frequently discussed advantage of additive manufacturing over other technologies is the ability to fabricate full assemblies of parts. This is easiest with polymer PBF due to the lack of support material, although it can be readily achieved with almost all additive processes through clever design. The principal consideration is the distance between separate parts, which will vary from machine to machine and is also influenced by part geometry, process parameters, and material. Table 7.8 provides minimum values that can be achieved with each of the key additive technologies for product designers, although testing machine capabilities and process parameters is essential. Achieving these

Table 7.8 Minimum distance between moving parts

Polymer PBF	FFF	Metal PBF
SLS = 0.3 mm MJF = 0.7 mm	0.5 mm	0.2 mm

minimums will also depend on the specific geometry, with large surface areas in proximity creating a build-up of heat that can cause unintended fusion, or difficulty in removing loose powder. In these situations, larger distances are needed between parts.

For additive technologies requiring support material like material extrusion and metal PBF, spacing between parts will also need to consider the removal of supports, either directly between parts or through attaching parts to the build plate. For material extrusion, minimising the need for any support is ideal and can be achieved by observing guidelines on overhang angles. Utilising dual extrusion systems can also simply the production of assemblies, with one filament being soluble. However, with metal PBF, the supports cannot be avoided due to the requirement to anchor parts to the build plate and transfer heat, so consideration must be given to print orientation and the placement of supports during design.

Colour 3D printing

The ability to produce functional multi-colour or full-colour parts is limited to polymers. Gypsum has been used for full-colour binder jet printing for many years, and sheet lamination processes can include colour printing. However, these are not common for product design, as they do not result in strong functional colour parts. Multi-colour printing simply refers to the ability to print in more than one colour and can be readily achieved with accessible FFF 3D printers featuring two or more extruders, or with a mechanism for swapping filaments during the print process. This can also be used for multi-material printing. Depending on the 3D printer and slicing software, multi-colour printing with FFF can be achieved in several ways:

- Separating a part into multiple pieces representing the different colour zones within CAD. These are then exported as separate files and set up as different colours (filament changes) during slicing.
- Designating faces or parts as different colours within CAD and exporting in a single colour file format, detailed in the next section. The number of colours will match the number of extruders or filament swaps available on the specific 3D printer.
- Manual programming of colour changing either within the slicing interface, or directly in the g-code. This is limited to changes at set layer heights.

Going forward, the more widespread availability of technologies that allow for complex colour patterns in end-use products from a product designer's point of view would be a welcome development. Colour application in CAD software may be as simple as applying one or several colours to entire parts or faces of a part, or as complex as wrapping a 3D form with a full-colour graphic – the options will be limited by the CAD software. Gaining the desired control may require the use of additional software outside of standard CAD packages, including software commonly used by games and animation designers (e.g., Blender), or advanced additive manufacturing software like Materialise 3-Matic. To 3D print in full colour, a file must be exported from CAD in a suitable format. Several different formats are useful and explained as follows:

- *OBJ (.obj)*: This format has historically been the most popular for colour 3D printing and, in simple terms, can be a similar mesh to that of a STL file but with additional colour information. This colour information is saved as a separate material template library file to the OBJ (with the .mtl extension) which includes details like transparency, reflectivity, and other material settings, as well as a separate image file (in PNG or JPG formats). These three files are all needed for colour 3D printing.
- *VRML (Virtual Reality Modelling Language, .wrl) and X3D (.x3d)*: Originally developed for web applications, VRML and the newer X3D files are largely similar to the OBJ format. However, only a 3D geometry file and an image (PNG) file are saved. The preference for OBJ or VRML may depend on available software.
- *3MF (3D Manufacturing Format, .3mf) or AMF (Additive Manufacturing File, .amf)*: As described in Chapter 1, the 3MF and AMF file formats have been specifically developed for additive manufacturing. They overcome the limitations of the previous file types by containing all information for 3D printing within a single file; this includes colours and materials, as well as process parameters for 3D printing and other information. These are increasingly available as a file export option in CAD software but may require newer versions/updates to be supported.

Without considering the science of colour, there are visible differences between the colours as seen on a computer monitor and what comes off a printer, whether it is a 2D or 3D printer. Full-colour 3D printing is no different, and research has shown significant differences in the outcome of printed parts compared to the source images, including the effects of post-processing treatments [3, 8]. Just as designers will calibrate a monitor and use colour profiles to improve the translation of screen to 2D print, calibration, or at least an understanding of colour translation, is necessary with full-colour printing. As shown in Figure 7.3,

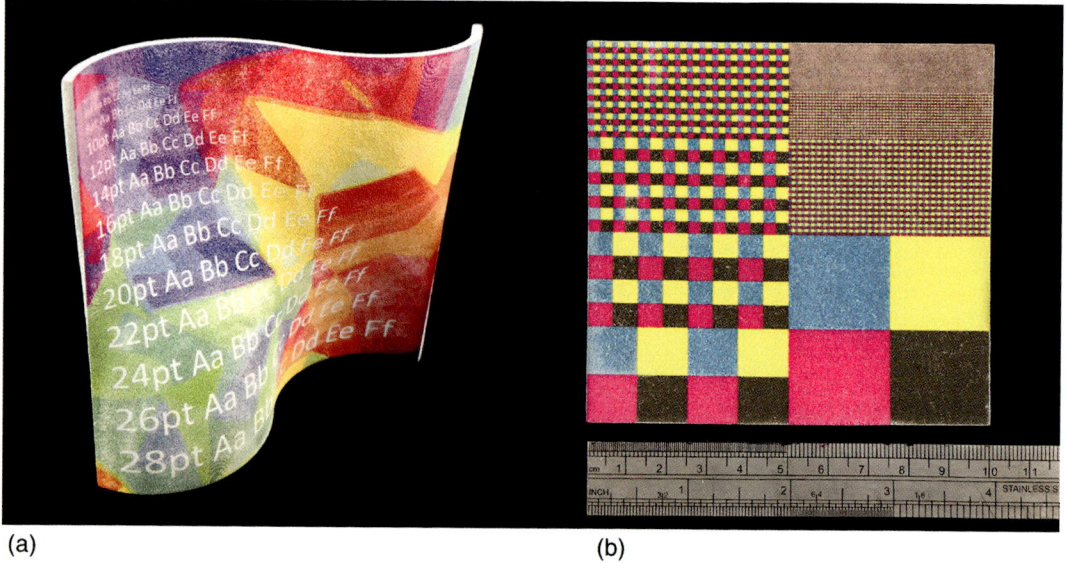

(a) (b)

7.3
Colour test pieces
printed on a HP Jet
Fusion 580 machine
prior to surface
treatment: (a) font
and colour testing
on curved geometry,
(b) CMYK swatches
which can be
sampled through
spectrophotometry
and compared to
digital colours.

test prints can provide a useful reference for colour 3D printing, including the clarity of graphic details like text, and are recommended when working with any colour technology. This is less of an issue with multi-colour printing with FFF, as the colours are not linked to the colours applied on screen and will look the same as the filament being fed into the printer.

PRE-PROCESSING AFFECTING DESIGN

Pre-processing generally refers to the steps between design of a 3D model and the physical 3D printing of that model, including the exporting of a suitable file for 3D printing, error detection/correction, slicing, and machine setup (aka process parameters). Decisions made during pre-processing will affect functional and aesthetic outcomes of parts, and therefore, they must be integrated into earlier design considerations. Hands-on experience with the range of pre-processing activities for additive manufacturing are essential for designers to understand the impact of these upon design, and to confidently specify these when handing a design over to a manufacturer or service bureau.

File exporting and mesh resolution

After a 3D CAD model has been created or sourced, the first step towards 3D printing is to export in a suitable file type. This sounds simple and is a standard part of going from design to any manufacturing process, but there are several options that can significantly affect the quality of the final product.

The principal factor for the most common STL 3D printing file format is the quality of the mesh. Most products are designed within CAD systems that provide mathematically accurate surfaces and edges (known as NURBS modelling),

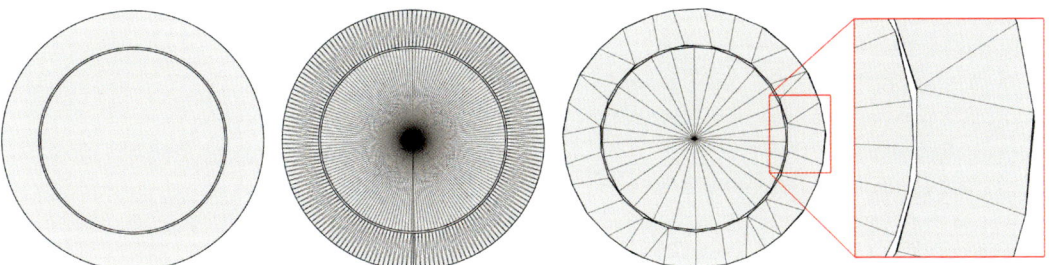

7.4
From left to right: Original CAD model; high-resolution STL mesh; low-resolution STL mesh; detail showing the variable spacing between assembled parts and faceted curves.

which must then be converted into a mesh. This mesh strips away all data, history, and other information used to create the part and simply represents the surfaces of a part as a mesh of unordered triangular surfaces. It is a bit like taking a perfect piece of vector-based art and converting it into pixels. The size and accuracy of these triangles in relation to the original design can be controlled upon export, and if the size is too large, they will result in visible flat triangular features when printed. This also affects the tolerance for critical features like the distance between assembled parts, as shown in Figure 7.4, and can cause unintended fusion between parts despite adhering to minimum gap tolerances. This will also affect the ability to print thin wall features [9].

Exporting with high resolution is key to mitigating these issues; however, it may not always be possible. File size is proportional to the number of triangles in a mesh, and complex geometries – for example, a lattice structure – can quickly increase mesh file sizes into hundreds of megabytes, or even gigabytes. This becomes difficult to work with for slicing and may not even be possible depending on software or computer hardware limitations. Striking a balance between surface quality and file size may be necessary, and additional additive manufacturing software for editing and optimising the arrangement and size of triangles in a mesh can be valuable.

Part orientation

The way a part is oriented for printing should not be random and can have significant effects on part strength, surface quality, print time, and post-processing requirements. This topic is worthy of a dedicated book, but for the purposes of this section, the focus is on why and how orientation should be considered during design, rather than left to moments before 3D printing. The following section on supports is also tightly linked to part orientation and should be considered simultaneously.

From a mechanical performance perspective, the layer-by-layer nature of additive technologies generally results in parts with anisotropic properties, meaning that the material performs differently in different axes based on print orientation [4]. For most processes, this means the weakest orientation is in the Z axis, with layers easily delaminated under loads compared to X or Y orientations. Parts requiring maximum strength with minimum material should be oriented with this in mind. During design, print orientation is related to the previous discussion of build volumes; the dimensions of the build volume with the desired layer properties will constrain the size of a part. For many SLS print volumes, the stronger X and Y orientations are often shorter than the weaker Z orientation, limiting the size of a part that needs to take advantage of material strength. To overcome this, a part could be designed with thicker walls, lattice beam diameters, and other details to ensure performance when printed in a weaker orientation, but this may not always be appropriate and will add print time and cost due to the increased material use.

In terms of aesthetics and the quality of surfaces, print orientation will also have a visible effect [10]. Class A surfaces (or those which are visible and critical to aesthetics) and important details will need to be oriented appropriately depending on the 3D print technology. This includes shallow curved surfaces which are best aligned perpendicular to the layer direction; else the stair-stepped effect becomes exaggerated and may require significant post-processing. This may be at odds with the mechanical characteristics desired, or the dimensions of the build volume, and a compromise may need to be made between aesthetics and performance.

Combining surface quality with important details like embossed or debossed text requires further consideration, as these fine details may be better suited in line with the layers for some additive processes (e.g., SLS, where the laser can perfectly trace the outline), compared with perpendicular to the layers (e.g., material extrusion). Another simple trick a designer should consider is whether to emboss or deboss text or details. Embossing text means that it protrudes from the surface rather than being cut into it; but, if extensive sanding or other subtractive process of the surface will be done, then perhaps debossing (cutting it in) the text should be considered instead.

For these reasons, it is important for a designer to specify the intended print orientation when handing files over to a client/manufacturer or outsourcing production to a service bureau. Often a service bureau, who may print tens or hundreds of parts from different customers in a single build, will rely on algorithms to automatically orient parts within a build volume that are maximised for efficiency and profit, rather than functional or aesthetic concerns.

A final design feature which may be dependent on print orientation is large flat surfaces. With many print technologies, particularly where heat is involved, these are not well suited to being horizontally oriented in line with the layers due to the differential cooling of material that occurs across the surface after fusion [11]. This can cause warping, leading to failure of a printed part, or even failure of a complete build as an extrusion nozzle or powder roller collides with the warped material. Orienting a large flat surface vertically will improve printability; however, as outlined earlier, this may also have negative mechanical performance issues.

Supports

Synonymous with print orientation is the need for support material with most additive technologies, including material extrusion and metal PBF. Support material is a bit like scaffolding used during the construction of buildings; it is temporary and used to support overhanging sections of a part while it prints. Once complete, support material is removed, revealing the finished part. Polymer PBF and some binder jetting processes do not require any support since parts are suspended within the powder.

The way a part is oriented will directly influence whether a part requires support or not, or how much support and in which locations. Good design for a specific additive process will minimise the need for support material, which in turn reduces print times and costs as well as cuts waste and simplifies post-processing. Removing support material can be a laborious manual process requiring several hours for complex parts, particularly metals, and care must be taken to hide the attachment points between part and support. This amount of labour may be OK for a one-off prototype or product, but it is not a scalable long-term strategy. Designing to minimise support is a fundamental part of designing for additive manufacturing, just as other manufacturing processes have their own best practice processes implemented during design.

A simple way to illustrate the basic principles of support is known as "THEY," illustrated in Figure 7.5. This refers to the fact that the letters "THE" are constructed from horizontal features known as overhangs (in the case of T and E) and bridges (in the case of H) which are floating in space. Without support material, gravity causes the initial layers of each overhang to droop, which can eventually correct as layers continue but may cause more significant print failure. In contrast, 3D printing the letter Y demonstrates how angled sections can be successfully printed without issues. Some of the problems with "THE" may be solved through orientation of the elements, although as described previously, this comes with numerous considerations for the applications, including functional and aesthetic. Others can be designed out completely, but only through an understanding of the final print orientation during design.

The top row in Figure 7.5. shows "THEY" letters printed in vertical orientation without supports. The second row shows how a changed print orientation can successfully print the same letters without supports. The third row shows how the original orientation can be maintained by adding supports. The fourth row shows how the letters can be modified to be narrower to improve printability without requiring supports. The bottom row shows some more significant design changes to incorporate chamfers or fillets to remove the need for supports.

Generally, across additive technologies, an angle of $\geq 45°$ to the build plate can be safely printed without support material. It is possible to print lower angles without supports, but that will require careful modification of process parameters, including the reduction of printing speed and increased use of cooling to rapidly solidify the plastic before gravity can take effect. Slicing software will allow the desired angle where support material is required to be specified and will automatically place supports for a part based on this setting. Additional supports can be manually located where necessary or removed from areas where the automated algorithms have been overzealous in support placement.

For metal PBF, supports serve a dual function; anchoring the parts in place to minimise the risks of warping due to the stresses cause by rapid heating and cooling, and transfer of heat away from the part to further minimise warping. Therefore, even a well-designed part will require some support, although designers should consider placement of supports in areas not critical to aesthetics, and to ensure tools can easily access support locations to remove them. This may also require designers to work with technicians to manually place supports during slicing, rather than relying on automated algorithms which will not know critical surfaces or details where support should be avoided.

Packing density

For business managers, packing density may be one of the more important pre-processing considerations, as it can significantly vary the cost of operating

3D printers. The packing density is most commonly a consideration for PBF technologies and is the amount of sintered versus unsintered material within a particular build volume. Another way of looking at it is how optimised the layout of parts are within the build volume. This is typically measured as a percentage, with 0 percent representing an empty build volume, and 100 percent representing the entire build volume being sintered as a solid block. Typically, a packing density of 7–12 percent or higher is ideal, although this will be highly dependent on several factors including the geometry and quantity of parts, as well as the way costs are calculated for internal or external print jobs. Most PBF systems run on a mix of used and virgin powder, which varies from manufacturer to manufacturer. If the powder mix ratio is, for example, 65 percent used powder to 35 percent virgin powder, then the closer the packing density gets to 35 percent, the less powder that gets wasted.

Because of the time required to set up a print job and then wait for the build to cool (which may be six hours or longer, often the same duration as the print time), as well as post-processing and recovering powder, it is more cost-effective to fill a PBF build volume as much as possible rather than use a low packing density (nested parts ready for 3D printing shown in Figure 7.6). Because the fusion process of each layer is fast, the main factor determining print time is the number of layers, or height of the build. Basically, the taller the build, the longer it will take to print. For example, a single 200 mm tall part printed on its own (low packing density) will take a similar time to print as many smaller parts filled to a height of 200 mm with optimal packing density; the difference is that the many smaller parts will divide the cost of production, including pre- and post-processing, while the single part will bear the full cost on its own. This can be a deterrent for utilising the full diagonal space within a build volume, requiring the full build height to be run each time that part is needed.

7.6
Example of packing for the HP Jet Fusion 580 3D printer.

A useful technique employed by many businesses and universities is to fill unused areas of a build with small test pieces or promotional items to increase the packing density of every build. These essentially become free to print compared to leaving the space empty. For designers, packing density is valuable to consider during the design of a product, as being able to print more within a single build lowers the cost per product. As a general rule, 2 mm to 5 mm is the minimum distance that should be left between separate parts in a PBF build, although 10–20 mm is preferred around parts which have thicker areas of fused material. This can be much smaller for FFF or resin-based processes.

For material extrusion, packing density is less of a concern, as little time is saved by having two parts printed together versus separately, and parts must sit on the build plate rather than being able to freely fill a full 3D build volume. However, there are tricks to improve throughput; for example, stacking multiple parts together that can be sliced apart after printing. This strategy can be useful for running print jobs overnight, ensuring a full 8–12 hours is maximised rather than printing only one part requiring a couple of hours, leaving a machine idle for most of the night. The surface quality of contact surfaces will be reduced, but for many applications it will be suitable or can be cleaned up during post-processing. Designers can deliberately accommodate part stacking as a feature of part geometry, ensuring parts can sit atop each other with minimal support material, and even adding sacrificial features that anchor parts together in a stack, to be manually trimmed away after printing (similar to an injection-moulded sprue).

Layer thickness

Layer thickness is largely determined by hardware and material limitations, and it is often assumed that printing with the smallest layer thickness is the best option, as it will give the best surface finish. If a good surface finish is the main criteria, this may be true. However, for many applications, bigger is better, and designers are increasingly embracing the layered aesthetic of 3D printing as a benefit rather than a limitation (as detailed through several of the furniture case studies in Chapter 6, for example). If a part is made up mainly of flat vertical and horizontal faces, for example, a thicker layer thickness may be perfectly acceptable, whereas for a part with many curved surfaces, a thinner layer thickness may be better.

No matter the additive technology, layer thickness directly affects print time; a part printed with a 0.1 mm layer thickness will take twice as long as the same part printed with a 0.2 mm layer thickness. For FFF technologies, where a typical desktop machine may be capable of layer thicknesses ranging from 0.1–0.8 mm (depending on extruder capabilities), this can mean the difference between printing one part in 24 hours at a 0.1 mm layer height, to printing the same part in just 3 hours at a 0.8 mm layer height (or 8 parts in 24 hours). As well as affecting surface finish, layer thickness selection will also impact part strength, with research showing that parts printed with thicker layers are weaker than those with thin layers [12]. Thinner layers are ideal for many end-use applications outlined in Strategy 2 of this book and will provide the most accurate representations of CAD geometry. However, thicker layers can play a useful role in several applications:

- *Quick prototypes*: For iterative design, getting a part within hours instead of days is valuable. Surface finish doesn't need to be perfect for rough prototypes.
- *Jigs and fixtures*: Rough surfaces are typically fine for parts being used in manufacturing applications where functionality is far superior to aesthetics.
- *End-use products*: Where the layers are used as an aesthetic feature.
- *Time-critical applications*: Where products are needed in emergency or disaster situations – for example, during the COVID-19 shortages of personal protective equipment – or for temporary repair parts to be used until a replacement can be sourced, maximising the speed at which a solution is found can be critical.

Infill

A unique opportunity with 3D printing due to the layer-by-layer process is to modify the internal geometry of solid parts, or solid sections of parts. In FFF, this is commonly referred to as infill or, with PBF, as honeycombing. With many other manufacturing processes, solid sections must be kept to a minimum, as these can cause problems due to differential cooling (e.g., shrinkage) and add significant weight. Yet, by modifying the internal geometry of a solid area for 3D printing, weight can be kept low whilst giving the appearance of being solid. Infill is most seen with material extrusion printing, although it is popular with most other printing processes. For polymer and metal PBF as well as resin printing, consideration must be given to removing unfused powder or resin trapped within solid external walls – this can usually be achieved with several escape (or drainage) holes, although some material may remain trapped inside. In metal or polymer PBF, if weight is not an issue, it can still be beneficial to honeycomb the part and leave the unmelted powder trapped inside, as it can greatly reduce the print time and cost.

Figure 7.7 shows a collection of common infill geometries available with FFF 3D printing. These are automatically generated within slicing software and can be modified in terms of their density from 0 percent (hollow) to 100 percent (solid material). It is also possible to manually model any desired infill, either directly within CAD or in a secondary piece of software that may allow variable lattice structures or other geometries to fill solid zones.

The density of the infill, as well as the pattern (see Figure 7.7), will affect part strength and stiffness in different orientations [13]. Additionally, the higher the density of lattice, the longer the print time and the higher the cost due to increased material use. Considering these factors, it is important to design solid

sections of a part with intent; infill selection, whether implemented as part of the CAD process or at the latter slicing stage, is an important design decision that should not be left to a technician or automated process prior to 3D printing. In its extreme, infill generated during pre-processing can be used as a design tool to produce complex part aesthetics, just like a lattice structure – i.e., a part can be printed with no solid exterior skin but be completely printed as an infill structure.

POST-PROCESSING 3D PRINTED PARTS

The 3D prints featured in news stories or at trade shows can often give the illusion that 3D printed parts are smooth and glossy, vibrantly coloured, clear, or polished. It is not surprising that many are disappointed when they first experience a part coming out of a 3D printer. The reality is that achieving these finished outcomes almost always relies on post-processing tools and technologies, and often these vital tools are left out of initial purchasing budgets. Many of these tools are not even sold by the manufacturers or suppliers of 3D printers and are a combination of new and old techniques for finishing manufactured parts.

Parts produced through additive manufacturing can be treated in much the same way as any other manufacturing process. As a minimum, there will always be some manual clean-up of parts after printing, including removal of supports or loose powder. But this section will focus on some of the techniques for improving or modifying the exterior surfaces of parts after these minimum requirements have been completed.

Painting, powder coating, anodising, and graphics
Many of the most common and affordable additive processes can print in only a single material/colour at a time. To achieve a desired finish, painting, powder

coating, anodising, and the application of graphics are common, and the options will vary depending on the additive process and material. Painting is the most basic colour technique that can be applied to any polymer or metal process. Just like painting a vehicle or a figurine, the process of painting a 3D print will typically involve several stages of sanding, priming, painting, and clear coating. It is best to use a thin layer height for parts being painted so that a combination of sanding and filler/primer will more easily hide the layer steps. Painted 3D prints will be most common for prototypes and some one-off artisanal products, and designers need to accommodate the amount of material removed/added from sanding and painting during design, particularly areas requiring tight tolerancing. 3D printing of full assemblies may not be possible where a painted finish is required, and complex geometries like lattice structures will need to consider how sanding and painting process will reach all surfaces.

DESIGNER'S COMMENT

Avoid having cosmetic details or logos protruding from your 3D printed parts. It makes the part harder to sand down without damaging the cosmetic details.

Powder coating and anodising processes can also be applied to metal 3D prints in the same way as standard metal components. Powder coating involves spraying a powder which is electrostatically attracted to the surface of a part. When cured in an oven, the result is a uniform-coloured surface that is much stronger than a painted surface. Anodising is performed most on aluminium through an electrolytic process, growing a protective surface which can be dyed. Unlike conventionally manufactured parts, the layer-by-layer additive process can be leveraged to accommodate internal details to be used as suspension points for these coating processes, ensuring all visible surfaces are coated in one go. When anodising, it is always best to create sample colour swatches, as the alloys used in metal AM may differ from conventionally anodised alloys.

Some of the most eye-catching effects applied to 3D prints utilise hydrographics, also known as hydro dipping or water transfer printing (see Figure 6.16 in Chapter 6 for an example). This process involves printing a 2D graphic or image onto a soluble substrate which floats on the top of a water surface. After application of an activator solution to dissolve the substrate, a 3D print can be dipped through the floating graphic, stretching as it adheres to the 3D printed surfaces. This can adhere to polymer or metal prints, although limitations will apply to geometries suitable for this process; for example, lattice structures and other complex 3D geometries would not pick up the graphic on the interior surfaces or details. This process is best suited to parts with flowing, continuous surfaces where the application of the graphic can be controlled and reliably applied.

Dyeing

Dyeing is predominantly used with the polymer powder bed fusion process and leverages the inherent hygroscopy of nylon parts from this process to absorb ink. The polyamide used with selective laser sintering (SLS) results in white parts that can be dyed any desired colour using processes common in fashion and textiles to dye fabrics. Dye is added to warm water, and the parts simply soak in the solution for several minutes until the desired colour is achieved. This is a less labour-intensive process than painting that can even be automated, and it is commonly offered as an additional service by manufacturers and service bureaus. However, multi jet fusion (MJF) parts are naturally dark grey in colour due to the colour of ink used for the fusion process. This limits the range of colours that can be dyed, with parts commonly dyed black or not at all, although some aftermarket processes are emerging to increase colour options.

Vapour smoothing and epoxy coatings

Sanding away the layered finish of 3D prints to achieve smooth surfaces is a laborious and messy process, and there are other ways to improve the surface finish of polymer 3D prints that mitigate the need for sanding altogether. Vapour smoothing is one of the best-known options that utilises chemicals like acetone to dissolve the exterior surfaces of a print. Rather than soaking parts in solvent chemicals, which could aggressively dissolve the entire part in a matter of seconds, vapour smoothing involves heating a solvent so that small droplets become airborne within a confined area, which more gradually dissolves the exterior surface and blends the stepped layers. If left too long, the process will completely dissolve a part, so timing is critical.

It is important to note that vapour smoothing can be quite dangerous, as the chemicals used and the fumes produced during the process can cause breathing difficulties, skin damage, and other injuries, so caution is necessary. The appropriateness of this process is also reliant on chemical reactions between printed material and solvent, and different solvents will be necessary for different polymers. The most well-known is the use of acetone for ABS and ASA materials, and while some of the online do-it-yourself guides may sound like science experiments, there are commercial devices that can dramatically improve the safety and control of the vapour smoothing process.

One of the main points designers must consider if intending to use this process is the thickness of geometry; thin, fine details will quickly dissolve away while thicker, large surfaces will take more time to experience the effects of the solvent. Therefore, geometry should remain as uniform as possible and delicate details like embossed text should be avoided. It is also necessary to consider the thickness of material that is removed from this process and allow enough material for parts to remain functional. Positive findings from vapour smoothing are that it can slightly improve the surface hardness of parts, although the elastic modulus decreases [14].

Another method of smoothing that is less hazardous is the use of epoxy finishing products such as XTC-3D™. Such products typically involve mixing a two-part solution which can be brushed onto a 3D print, with the desirable property of self-levelling to fill the steps in 3D print layers. The result is a smooth, glossy surface which can be left as is or painted, although the clear coating will change the colour of the underlying surfaces, so it is best to test samples before committing this finish to a final product. Much like considerations around painting discussed earlier, products to be finished with an epoxy coating may need to limit the use of complex geometries like lattice structures or internal details where it may be difficult to brush the epoxy on all surfaces.

A similar opportunity that can be employed with SLA/DLP printed parts is to dip them into the same resin used for printing and cure them in a post-curing station. This is not a common treatment but can be particularly useful to achieve clear SLA parts without extensive sanding and polishing processes. Again, this will alter a part's dimensions, and controlling the flow of resin on and around part surfaces can be difficult.

Bead blasting and tumbling

Abrasive materials can help remove the stair-stepped layers on surfaces and clean up other details such as loose powder, support removal areas, sharp edges, and other surface defects. Bead blasting, also known as sand blasting, is a common process in manufacturing that can be used with 3D printed polymer and metal components. A blasting media such as sand, glass, or plastic is shot from a nozzle under high pressure, normally within the confines of an enclosed cabinet, and an operator either moves the nozzle around a part or moves the part around a fixed nozzle to remove material. This is a particularly common post-process for powder bed fusion parts, and used polymer powder that cannot be recycled back into a 3D printer can itself be used as blasting media for both SLS and MJF parts.

Depending on the combination of air pressure, blasting media, and time spent blasting a part, different surface effects can be achieved, ranging from simple removal of loose powder to complete surface finishing with a matte appearance (different surface finishes are shown in Figure 7.8). Bead blasting systems can be inexpensive desktop systems for infrequent small part processing, or they can be large automated systems to streamline production. The size and volume of parts will determine the appropriate type of bead blaster required within a manufacturing setting.

Tumbling is another common industrial process that is applied to 3D printed polymer and metal parts. Abrasive media, which may feature specially shaped geometries and different material, is rotated, or vibrated inside a tumbling machine. When parts are added, the abrasive media gradually removes loose material and sharp edges, smoothing the surfaces. This is an automated process that can be done in batches, simplifying part clean-up. The disadvantage compared to bead blasting is that the surfaces are treated uniformly,

(a)

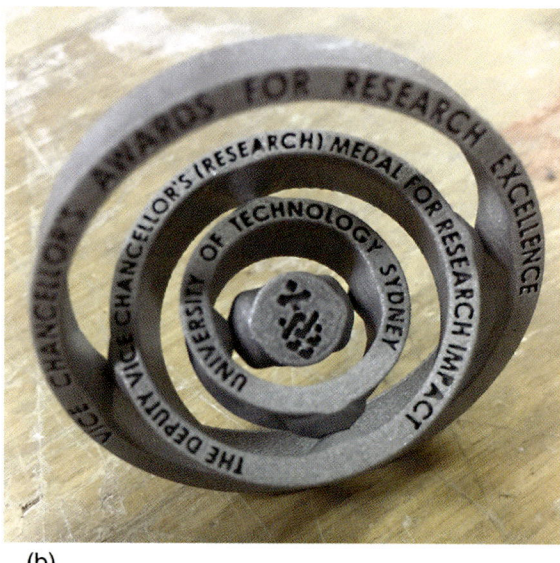

(b)

whereas in bead blasting, an operator can choose to focus on surfaces requiring particular attention (e.g., where support was removed), or to achieve a certain visual effect. In both processes, interior surfaces of complex part geometry may not be cleaned the same as external surfaces depending on the access for bead blasting, or the size of the abrasive media in tumbling. Where internal cavities required smooth or even polished surfaces – for example, internal cooling channels of a mould – abrasive flow machining can be used. This involves forcing an abrasive fluid through a part under pressure, gradually wearing away the surfaces.

7.8
Alternative surface finishes for metal prints – a part that has been polished (a), and the same part bead blasted with a more matte surface finish (b). (Novak)

Metal plating

Metal plating is a popular finish for both metal and plastic parts, and applying this process to 3D printed parts is the same as any other manufactured component. Electroless plating is the starting point for polymer parts and normally coats the part in a nickel or copper layer between 1 and 50 μm thick. After initial plating, almost any other metal plating can be applied over the top, which will increase the strength of polymer parts while keeping weight low [15]. Examples of plating materials include chrome, tin, platinum, gold, and silver. The plating will also improve corrosion and abrasion resistance. Metal plating of polymer parts can be significantly more cost-effective than 3D printing in metal, and due to the large sizes possible with polymer processes, it allows large metal-like parts to be produced in a single piece. End-use applications include jewellery, electronic circuits, antennas, vehicle components, furniture, consumer electronics, and many more. Plating can also be cost-effective for prototypes that will later be manufactured in metal to give the right look and feel to clients or customers.

Metal 3D printed components can be plated using electroless or electro-plating processes and may be done to improve aesthetics (e.g., gold plating) or resistance to chemicals or abrasion. Designers must consider the additional thickness to all surfaces receiving metal plating and the effect on critical toler-ance values. Much like powder coating and anodising, hidden details to allow for hooks to hold parts throughout the metal plating processes can help improve the results of plating. Other methods of metal plating that may be used for 3D printed parts include physical vapour deposition and vacuum metalizing.

Polishing, machining, and metal surface finishes

Parts printed in any metal process can be machined, polished, or given surface treatments just like any other metal manufactured part (post-processing stages in 3D printing are shown in Figure 7.9). This may occur before or after bead blasting or tumbling (described previously), depending on the desired finish. Polishing is one of the more popular surface finishes that can remove the tell-tale signs of being 3D printed. This is best done after bead blasting and can involve a staged approach of using different sizes/grades of abrasive media to achieve the desired polished appearance. Polishing of 3D prints is typically a manual process, especially due to the unique geometries of each part, and may rely on a mix of machine- and hand-based processes. This can be time-con-suming, especially for complex parts, and requires a skilled technician who can minimise the time required to achieve a desired finish and control the amount of material removed. Since polishing is a subtractive process, extra material is required for a design to accommodate the amount of material removed from the bead blasting and polishing processes; 0.5 mm should be enough extra material for most polishing applications. Some manufacturers now also supply high-speed centrifugal finishing systems that can speed up the polishing process.

Subtractive machining processes can be used both for aesthetic and func-tional finishing, and some 3D printers are capable of both additive and subtrac-tive processes within the same unit (known as hybrid manufacturing). However, it is more common for machining to occur separately using traditional

7.9
From left to right: Raw aluminium ring from SLM process with some support material removed; ring after support removal, filing, and bead blasting; ring after polishing for five minutes with a polishing wheel (Novak).

equipment, including manually operated or computer numerically controlled (CNC) machines like mills and lathes. Machining can be used to achieve a fully polished part, or only for critical features such as holes or mounting surfaces which need to be accurate. This includes surfaces for moulding applications. Holes should be printed in the vertical orientation where possible to improve their accuracy and roundness.

As with polishing, designers must accommodate the material removed by increasing the thickness of parts during design by 0.5–1.0 mm. Additionally, including anchor points or alignment geometry to parts can dramatically simplify the process of setting up a part for machining, particularly for CNC machining processes where the machine must know exactly where the part is to accurately machine it. Counterintuitively, it can sometimes be best to put supports on the surface where the best surface quality is needed, because to achieve that best surface quality, it will likely be machined to finish irrespective of the evidence of support removal.

Shot peening will also modify the visual appearance of metal parts, with the added benefit of improving resistance to cracking. Round metallic, glass, or ceramic balls are blasted with compressed air at a part, like bead blasting, except that rather than being abrasive and removing material, the exterior surface undergoes plastic deformation, flattening rough textures and details while also producing beneficial compressive stresses. It is like thousands of small hammer strikes across the surfaces of a part. This may be used as an alternative to bead blasting as the first post-process after support removal.

Heat treatments

The extreme temperature fluctuations during metal PBF, particularly DMLS/SLM, result in the build-up of residual stresses, and while support material will help resist deformation of parts, these stresses can cause parts to deform or crack later during use. With additive manufacturing – particularly in metal powder bed processes, excess material that does not add real functional value to the product is the most critical factor to reduce in improving the design. This is because large masses of material not only greatly increase print time and cost but also are one of the main contributors to residual stress. Like any metal process, heat treatments can be used to relieve internal stresses once a part has been manufactured and is common practice immediately after powder removal while parts are still mounted to the build plate. This is typically done in a furnace, and it may also be done within a vacuum environment, with the temperatures and timing highly dependent on the material.

Additional heat treatments can be used to modify metal part properties, with Hot Isostatic Pressing (HIP) often used to improve part density [16]. While metal PBF can theoretically achieve densities up to 99.9 percent, this can be affected by powder size consistency and process parameters, and the actual density of a part may be lower, exhibiting a small amount of porosity. HIP involves heating the part within a high-pressure, gas-filled chamber which plastically deforms the part uniformly from all directions, compressing pores and achieving 100 percent density. Annealing is another heat treatment which can be used to improve ductility of metal parts like aluminium and stainless steel. It is also possible to anneal polymer 3D prints, with evidence that annealing improves the mechanical strength of parts by relieving internal stresses and improving adhesion between layers [17].

DESIGNER'S COMMENT

The most challenging part of AM is that everything is controllable: porosity, surface finish, hardness, and so on. There is often a compromise between quality parameters; i.e., making a part denser may reduce surface quality. So, understanding the various parameters that influence your part quality are essential. Often the real question is: What is good enough?

MAXIMISING PRODUCT CHOICES

Most product designers have been using rapid prototyping in one form or another for quite a long time. This might be desktop 3D printing, or some of the more sophisticated 3D printers from service bureaus that enable prototypes that are suitable not just for visual testing, but also for functional testing. The criteria discussed here provide product designers with a basis for selecting applications and for design detailing for additive manufacturing technologies. These criteria involve consideration of the impact on business practice and business models, including supply chains, because the reality is that it is not a straightforward decision. For example, depending on the additive manufacturing technology, the machines can be expensive to buy. For metal additive manufacturing, for instance, prices can range from US$530,000/450,000 pounds sterling up to US$1.5 million/1.3 million pounds sterling, depending what kind of machine is selected and what build space is needed.

The machines can be challenging to install as well as to run and maintain. It is not that straightforward to find qualified, experienced technicians who have knowledge about additive manufacturing. Getting the desired output is also not straightforward, even though it can sometimes be presented as if it will be. This includes the design detailing for the specific technology (illustrated in Figure 7.10), but also working effectively with the range of settings on the machine to maximise the output. It can also be a problem if the technician is inexperienced in using the settings beyond the basics and able to adapt the settings to maximise the output. Working with additive manufacturing is not like working on an additional conventional machine, as frequently, to use it properly and exploit the opportunities it provides, it requires a change in business practice – and without a change in business practice, it can be a costly mistake.

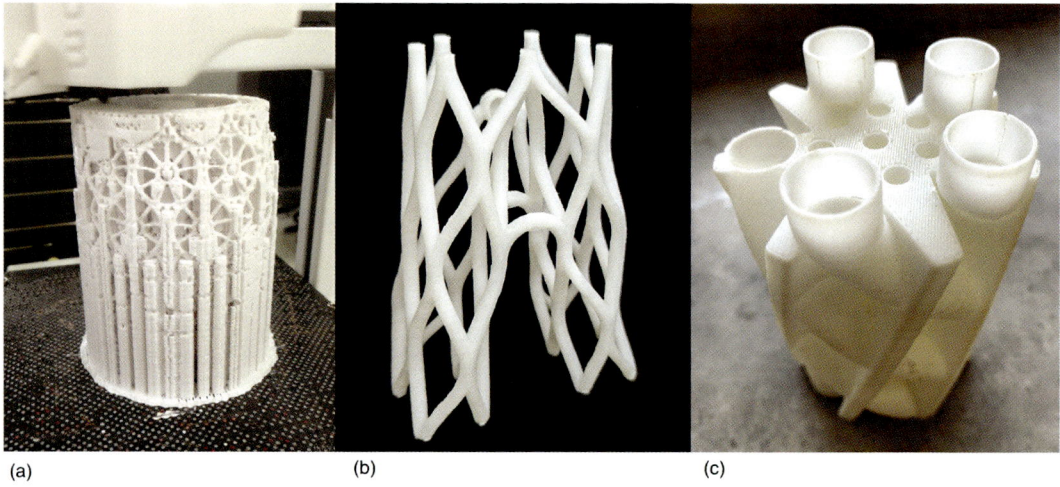

(a)   (b)   (c)

7.10
Test pieces: Support structure for FFF can be difficult to remove (a); designing to eliminate support structure (b) (Novak); complex structures avoiding supports (Matt Deschon).

DESIGNER'S COMMENT

The resident evil in additive manufacturing is support material. Support material is temporary sacrificial material that is used while the part is being printed. This must be removed and disposed of once the print is finished. Support material takes time to print and time to remove, and it leaves an inferior surface finish. To minimise the use of supports, consider replacing temporary supports with permanent walls in your designs. These walls can be solid, latticed, or perforated. Consider changing the angles of features to make them self-supporting.

Developing a part starts with the fundamentals of what the part is required to do, then additive manufacturing is used to map out the ideal geometries and expand the structure, exploiting the capabilities of the technology. Here the reduction in weight is measurable and potentially a benefit to the client. Other potential benefits are less easily quantified; for example, where multiple parts are being consolidated into one, reducing potential weak spots where components are joined, as well as the cost of joining those parts or potentially repairing or replacing parts in the future.

Understanding the complex interrelated impacts of the different products created with additive manufacturing is key to maximising product choices with the technology. Designs that maximise the opportunities that additive manufacturing offers make a part more viable. Alongside these considerations, product designers need to have a good understanding of design for process. Again, this is not straightforward, as clients tend to refer to additive manufacturing as if it is a single technology, when in fact there are seven defined families, plus hybrids, that, between them, have over 40 quite distinctive types of additive manufacturing. These all come with their own requirements, their own constraints and opportunities, and their own set of design detailing rules. These processes can be quite different, so within material jetting, for example, you can have multi-material and/or multi-coloured processes, and then you can have more basic dual-material processes that involve a resin and wax support material – very different applications, and different requirements in the way designs are created for them and their use. Therefore, it is critical that, when discussing additive manufacturing, the actual process is clarified.

There is frequently confusion between what can be printed with one technology and what cannot be printed with another. Designs have to be developed for the specific process that is appropriate for that design, and industrial clients have to have a good understanding of these differences and their implications for production.

TO 3D PRINT OR NOT TO 3D PRINT?

So, with all the design details and practical requirements, is it worth it? Is additive manufacturing a worthwhile pursuit? That is a complicated question to answer for a company. For a country, wholesale technology adoption of additive manufacturing would enable greater agility in manufacturing and added value to products. From a logistical and sustainability point of view, therefore, there appear to be benefits, especially when viewed through the current lens of supply chain disruption [18]. But what are the implications for an individual company? Is it profitable, and will it be disruptive? Without question, additive manufacturing has the potential for added value products for most companies, if it is used correctly [19].

In summary, if end-use additive manufacturing of parts is new to a company, the starting point should be: If it can be made using conventional production, steer clear! If, however, the parts benefit from increased geometrical complexity, especially where lattice structures are involved, or a consolidation of parts, or the outcomes open new markets, they have greater potential to be commercially viable. The question is, does the part being proposed exploit the characteristics and opportunities of additive manufacturing enough to add sufficient value, to make the technology adoption commercially viable, especially with the more expensive machines? Can it be designed for additive manufacturing, for example, to reduce post processing, especially in metal parts? But in each case, what would be the impact on existing business model and supply chain practices – good and bad – and what are the costs – financial, practical, and managerial – and potential profits or benefits for the business.

Overall, manufacturing as a sector needs to be fully informed on the complexities of the technology and products appropriate to different additive manufacturing technologies for different manufacturing situations, so that in working to advance manufacturing, designers can support the exploitation of additive manufacturing technology. Beyond the technical and design aspects of using the technology, businesses need to work with digital business innovation expertise to make informed decisions on if, when, and how it is relevant to them.

REFERENCES

1. Diegel O, Nordin A, Motte D. *A Practical Guide to Design for Additive Manufacturing*. Berlin: Springer; 2020.
2. (ISO). *Additive Manufacturing – General Principles - Fundamentals and Vocabulary*. Switzerland: ISO – International Organization for Standardization; 2021.
3. Badar F, Dean LT, Loy J, Redmond M, Vandi LJ, Novak JI. Preliminary Color Characterization of HP Multi Jet Fusion Additive Manufacturing with Different Orientations and Surface Finish. *Rapid Prototyping Journal*; 2022.
4. Cai C, Tey WS, Chen J, Zhu W, Liu X, Liu T, et al. Comparative Study on 3D Printing of Polyamide 12 by Selective Laser Sintering and Multi Jet Fusion. *Journal of Materials Processing Technology* 2021;288:1–12.
5. Rael R, Fratello VS. *Printing Architecture: Innovative Recipes for 3D Printing*. New York: Princeton Architectural Press; 2018.
6. Novak JI, O'Neill J. A Design for Additive Manufacturing Case Study: Fingerprint Stool on a BigRep ONE. *Rapid Prototyping Journal* 2019;25(6):1069–1079.
7. du Plessis A, Yadroitsava I, Yadroitsev I, le Roux SG, Blaine DC. Numerical Comparison of Lattice Unit Cell Designs for Medical Implants by Additive Manufacturing. *Virtual and Physical Prototyping* 2018;13(4):266–281.
8. Wei X, Zeng L, Pei Z, editors. *Experimental Investigation of PolyJet 3D Printing Process: Effects of Finish Type and Material Color on Color Appearance*. ASME 2019 International Mechanical Engineering Congress and Exposition; 2019; Salt Lake City, UT: ASME.
9. Novak JI, Liu MZ-E, Loy J. Designing Thin 2.5D Parts Optimized for Fused Deposition Modeling. In: Kumar K, Zindani D, Davim JP, editors. *Additive Manufacturing Technologies from an Optimization Perspective*. Hershey, PA: IGI Global; 2019. pp. 134–164.
10. Delfs P, Tows M, Schmid HJ. Optimized Build Orientation of Additive Manufactured Parts for Improved Surface Quality and Build Time. *Additive Manufacturing* 2016;12:314–320.
11. Armillotta A, Bellotti M, Cavallaro M. Warpage of FDM Parts: Experimental Tests and Analytic Model. *Robotics and Computer-Integrated Manufacturing* 2018;50:140–152.
12. Shubham P, Sikidar A, Chand T. The Influence of Layer Thickness on Mechanical Properties of the 3D Printed ABS Polymer by Fused Deposition Modeling. *Key Engineering Materials* 2016;706:63–67.
13. Dave HK, Patadiya NH, Prajapati AR, Rajpurohit SR. Effect of Infill Pattern and Infill Density at Varying Part Orientation on Tensile Properties of Fused Deposition Modeling-Printed Poly-Lactic Acid Part. *Proceedings of the Institution of Mechanical Engineers, Part C: Journal of Mechanical Engineering Science* 2019; 235(10):1811–1827.
14. Neff C, Trapuzzano M, Crane NB. Impact of Vapor Polishing on Surface Quality and Mechanical Properties of Extruded ABS. *Rapid Prototyping Journal* 2018;24(2):501–508.
15. Praveen BM. Electroplating of 3D-Printed Components. In: Muralidhara HB, Banerjee S, editors. *3D Printing Technology and Its Diverse Applications*. New York: Apple Academic Press; 2021.

16. du Plessis A, Macdonald E. Hot Isostatic Pressing in Metal Additive Manufacturing: X-Ray Tomography Reveals Details of Pore Closure. *Additive Manufacturing* 2020;34:101191.

17. Rangisetty S, Peel LD, editors. *The Effect of Infill Patterns and Annealing on Mechanical Properties of Additively Manufactured Thermoplastic Composites. ASME 2017 Conference on Smart Materials, Adaptive Structures and Intelligent Systems*; 2017; Snowbird, UT: ASME.

18. Sheffi Y. *The New (Ab) Normal: Reshaping Business and Supply Chain Strategy Beyond Covid 19*, British Columbia: Transoft Inc.; 2020.

19. Khorasani M, Loy, J, Ghasemi AH, Sharabian E, Leary M, Mirafzal H, Cochrane P, Rolfe B, Gibson I. A Review of Industry 4.0 and Additive Manufacturing Synergy. *Rapid Prototyping Journal* 2022;28(8):1462–1475.

8 3D printing sustainability and digital ecosystems

For product designers, creating practical and commercial product designs using additive manufacturing to directly respond to the growing sustainability imperative is an interesting challenge. One of the key problems in embedding sustainability in any product proposal is that, as with additive manufacturing itself, it is a complex field that covers a multitude of different ideas and approaches, technologies, and focus areas. When addressing sustainability, design strategies can appear contradictory, with several questions arising. For example, is the ideal to re-use materials or to design out waste earlier in the supply stream? Should the design be for longevity, or should the product designer opt for biodegradable materials and a "temporary product" approach? Similar to working with additive manufacturing technologies, separating out the different approaches to sustainable practice, and selecting which element to work on at any one time, can help the product designer be clear about the focus of the work for an individual project. This limits the sustainability ambitions of the project to more realistically achievable outcomes.

As with the three strategies outlined in Chapters 3, 4, and 5 of this book, integrating additive manufacturing technology into production practice requires that product designers develop experiences which enable them to build a system of different sustainability strategies that work for different contexts. When tackling additive manufacturing integrated with sustainability, it helps to be clear on the specific sustainability approach being used, how it is defined, and how the value it offers can be measured. This chapter considers eco-pluralistic sustainable design approaches, viewed through the lens of additive manufacturing. The aim is to provide product designers with guidelines to help differentiate sustainable product design using conventional technologies, and the opportunities for sustainability provided by additive manufacturing.

Clarifying which approach is being used in sustainable design with additive manufacturing can also help control the narrative around that and suggest appropriate measuring tools for considering how effective the solution may be in a specific use case. Product designers, clients, and consumers could all benefit from guidelines on evaluating the value of products designed using emerging digital technologies. The sustainability imperative is a critical element of current product evaluation and is likely to become more critical over time.

DOI: 10.4324/9781003122203-9

However, much as with understanding the complexities of lifecycle assessment for products and services that have grown over the last 20 years, it is not a straightforward exercise.

Commercially, additive manufacturing tends to be more viable when the production context and the full lifecycle of the product is factored into the evaluation anyway. This supports its consideration during a focus on sustainability, as this evaluation is fundamental to understanding environmental and societal impacts too. As additive manufacturing is becoming more viable for product designers at a time when the sustainability imperative is being universally recognised, additive manufacturing may be well placed to evolve with sustainability embedded throughout its development in the product design discipline. Dissimilarly, its basis in engineering began prior to the current sustainability imperative. Nevertheless, this potential in product design will be realised only if the discipline significantly builds its body of knowledge on the complexities of additive manufacturing technology, its applications, and its impact on business practice. Differentiation between sustainable design strategies and the ability to work with, and discuss, varying approaches as part of additive manufacturing optimisation needs to be sufficiently advanced to develop a defensible integrity to practice. To do this, product design academics, as well as professional product design associations, will need to show leadership in research and education.

8.1
Hexa-Phone
Amplifier won the
iMaterialise 3D
Printed Wood
Challenge. It is 3D
printed using
selective laser
sintering, with the
bulk of the powder
made up of waste
wood product
(Novak).

PRODUCT SERVICE SYSTEMS

A product service systems approach is becoming critical to design thinking for product designers from a sustainability point of view. The underlying principle is that the production, distribution, use, and renewal of products are viewed holistically. For sustainability, this means the ability to repair or recycle and to produce new items from reclaimed material. Examples for additive manufacturing

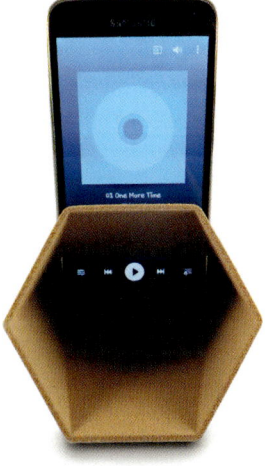

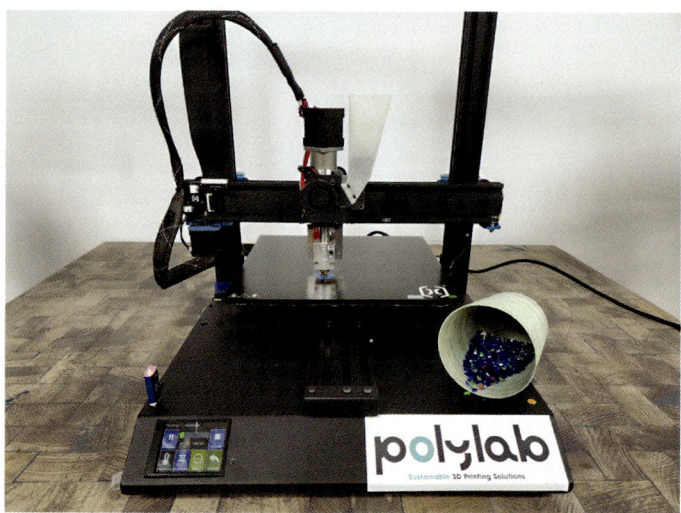

8.2
Australian start-up
Polylab developed a
screw-based
extruder system
which can utilise
shredded plastic
from discarded
prints, or other
waste products,
through a hopper
system. In this
example, HDPE
material is being
turned into a new
product. (Image
courtesy of Chris
Larkin, Polylab.)

include the use of sawdust depicted in Figure 8.1 and the recycling of plastic using a hopper, as in the Polylab example in Figure 8.2.

On a larger scale, these should be built into a system that connects the manufacturer and the consumer. Hawken provided a manifesto on changing practice in his 2017 publication *Drawdown: The Most Comprehensive Plan Ever Proposed to Roll Back Global Warming* [1]. As the "right to repair" policy discussions in Europe indicate, there has been a change in thinking about maintaining the longevity of products through repair, compared to the commercial built-in obsolescence of the last 50 years. Visionary company Michelin Tyres suggest that it could be possible, in the future, to renew tyres through the addition of 3D printed tread to the wheel. The use of cold spray additive manufacturing techniques is also being explored for repair applications, suggesting the possibility that additive manufacturing could be integrated into a repair-and-renew approach to boost sustainable practice. Another approach to designing for longevity is to create products that are flexible in use; for example, imagine a building with movable internal walls.

Products designed to be adapted into different configurations can meet sustainability needs. Product designer David Trubridge produces lights made from a myriad of small CNC parts that can be reconfigured to create different forms. Additive manufacturing could be useful with this approach in the design of connectors or building attachments to existing products.

TEMPORARY PRODUCTS

A different approach to sustainability is to accept that there is a demand, and at times a need, for products that can, and should, fulfil a short-term function. For example, creating an exhibition stand from materials that would last a lifetime would be an over-engineered waste. Even where a product is not designed for limited use, this approach recognises that there is a human desire for change.

Rather than work against this, the approach accepts that reality and tries to work with it. Examples from conventional product design include exhibition stands made from recycled cardboard, the use of lights projected onto buildings to create an atmosphere for an event (in place of custom-built physical installations), and biodegradable materials used for temporary products (including golf tees). Replacement parts form a popular area of 3D printing. It may often take considerable time to get a spare part for a machine, and if this means the machine cannot function, this can equate to substantial costs to the company. The ability to print a replacement part within a few hours can prove to be invaluable. Parts get printed both as permanent replacements, or as temporary replacements which last until the regular replacement part can be sourced.

For most consumer products, the reality is that they have a limited lifespan and should be designed appropriately. However, frequently they are not. There has been increasing recognition of this issue, as well as the increasing understanding of the need to legislate to ensure that products are not sent out of the factory without regard to where they will end up at their end of life. In Europe, legislation called "Extended Producer Responsibility" ensures that a manufacturer is responsible for products made by that company at the end of use. Once the product is no longer functional, the idea is that it is returned to the manufacturer for disposal. The manufacturer is responsible for any costs involved in disposal, subject to guidelines on what, and how, materials can be disposed. This legislation has had a significant impact of manufacturers, as it models the lifecycle of a product as circular, and designs have been adapted consequently. Cars, for example, need to be designed for disassembly as far as possible such that the materials used in manufacture can be reclaimed. This creates very different rules for product designers. Unfortunately, this legislation has had limited introduction into the United States and other countries, although it is spreading. It does demonstrate how changes in policy can directly impact the development of sustainability solutions for all governments.

If this legislation is viewed through an additive manufacturing lens, then it highlights the need for greater material reclamation. As filaments can be made using relatively low-tech extrusion equipment, and pellets can be directly 3D printed with robotic arms and even some desktop FFF machines, it could be possible to design products to be updated by the user by recycling the original materials into something new. Simultaneously, in line with the drive for "Right to Repair" legislation, again coming out of Europe and the UK, product designers can use additive manufacturing as the basis for designs that can be adapted by the user, because of the accessibility of low-end, digital fabrication tools. Even where an individual may not want to 3D print elements of a product to repair or update it, as shown in Figure 8.3, online additive manufacturing service providers can meet that need.

Material use in 3D printing is part of lifecycle assessment. One of the challenges that additive manufacturing has faced centres on perceptions regarding sustainability and sustainable practice. There is a strong view that the parts produced

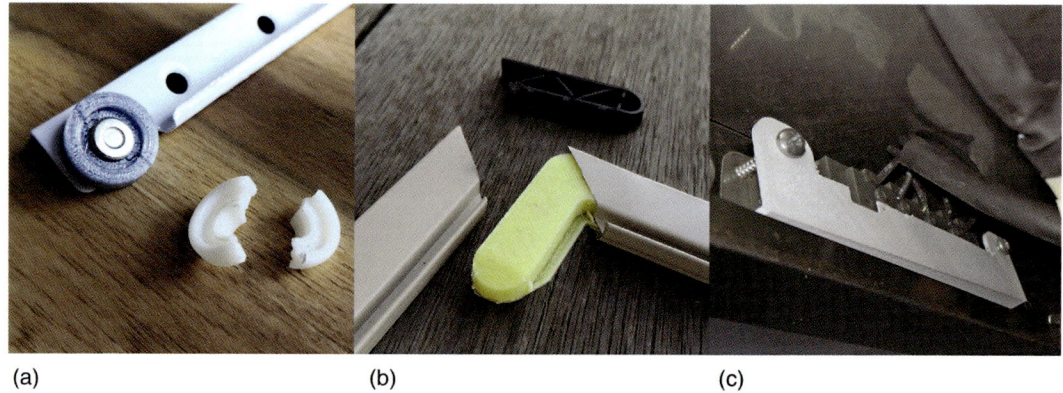

(a) (b) (c)

are predominantly plastic, despite a wide range of arguably more sustainable materials that can be used. Nevertheless, these also come with their own compromises; for example, wood powder having to be held together with glue, much as in medium-density fibre board. Equally, multi-material printers, such as the material jetting 3D printers used in anatomical models, suggest a tantalising future where products can be made from multiple materials with different characteristics within a single print. However, from a recycling point of view, this would cause similar problems to overmoulding in injection moulding, as the different layers could not be separated for recycling, even assuming the materials could be recycled.

With the growth in lifecycle assessment tools over the last two decades, and the growing use of embedded sensors in products as part of the approach termed the Internet of Things, the embodied energy in products that are evolved by the customer needs to become part of its lifecycle inventory. As the "digital threads" tracking parts, processes, and use of a product develop, they present a very interesting area of development for a more widespread approach to integrating sustainable additive manufacturing into the community. Addressing these complex challenges requires product designers who have a good understanding of additive manufacturing technology to start to build a body of knowledge on sustainable product design using additive manufacturing. The aim would be to exploit the additional opportunities it provides. As an industry, additive manufacturing providers need to continue to mitigate the inherent issues that undermine the technology's sustainability credentials. Working with product designers on design approaches could help this process.

8.3
Examples of 3D printing used to repair common household products: a replacement drawer roller (a), a corner joint for an aluminium window screen (b), and a bracket to retain control buttons inside a kitchen rangehood (c) (Novak).

MAXIMISING THE MATERIAL

In the book *Digital Eco-Sense: Sustainability and ICT – A New Terrain for Innovation*, Ryan [2] argued that greater investment is needed to be made in technology to improve the value gained from raw materials. 3D printing, particularly when utilised within a suite of other digital technologies such as 3D scanning technology, can enable high-value products to be realised from the materials available. There is inherently a significant raw material saving in the use of 3D printing over conventional production practices. Depending on the geometry of a part being made, a significant proportion of the raw material can be wasted during fabrication because of the constraints involved in using the traditional subtractive technique of milling. With 3D printing, if the part has been designed specifically for the technology, to reduce or preferably eliminate support structures, the value extracted from that raw material can be significantly increased. Light-weighting creates functional value-addition for products, but it also establishes a culture of material maximisation.

By using a shell and lattice-fill approach, components can be produced in metal additive manufacturing that add substantial value because of their reduced weight. The material reduction also substantially decreases printing costs. If a broad shift in approach occurred in manufacturing, where material value was made artificially high to change practice, for example, the investment in products would be directed away from material. This investment could then be channelled towards skills input to create value-added products. The pre-consumer waste generated during material management and product manufacture would be reduced, and preferably avoided, if the value of the material itself was much higher. From a sustainability point of view, the investments made in creating products with a high degree of complexity to meet challenging customer needs or market preferences, typical in additive manufacturing, would meet the strategy of creating added value from the raw material.

The geometrical complexities and light-weighting enabled by additive manufacturing allow product designers to explore more options based on biomimicry. Biomimicry more broadly is increasingly advocated for as critical to the survival of humanity and protection of the environment. However, it is challenging to integrate a biomimicry approach into consumer products. Janine Benyus is a leading advocate of biomimicry, providing information, publications, and workshop collaborations between scientists and designers [3]. The intent is to work with the environment rather than against it.

Product designer Ross Lovegrove uses 3D printing to mimic organic forms, such as bone, to light-weight products and reduce material use. Although conventional polymers and metals are mainly used in additive manufacturing, it is possible to use natural materials such as salt or coffee grounds for 3D printing, as demonstrated by architectural firm Emerging Objects [4]. There is considerable research on the use of natural materials such as shells and seaweed for 3D printing by organisations such as the SCION research facility in New Zealand. These materials are becoming more accessible, particularly in filament form available for desktop filament 3D printers.

INVESTED OBJECTS

Another approach to sustainability is to focus on the design of "Invested" objects. As the term suggests, these are simply products that an individual becomes invested in for one reason or another. If they then feel a connection

8.4
3D printed versions of "inflatables" artwork by Gregor Kregar, showing support structure on metal 3D print (Diegel).

with the object, they are more likely to keep it. Think of a museum piece: The value is rarely in the physicality of the object, but more in its associations, i.e., its cultural value. For an individual, a gift or homemade item can have an intrinsic value beyond a material sense. Stuart Walker describes this phenomenon in his book *Sustainable by Design* [5] as "enduring artefacts," where objects have an inspirational/spiritual or social/positional element that gives them value either to an individual or a community, as depicted in Figure 8.4.

In conventional design practice, the addition of a designer name to an object will add to its intrinsic value – for example, a Tom Dixon lamp or a Konstantin Grcic chair [6]. As additive manufacturing is relatively new to product designers when used for end-use products, there are fewer examples of high-profile names attached to items. An exception is Ron Arad glasses frames, which were exhibited in the Design Museum in the UK. However, several product designers are now emerging as leaders in design for additive manufacturing, and their products are becoming invested items as the profile of the technology, and the designers, grows. Assa Ashuach, for example, has been a disciplinary leader for the innovative use of additive manufacturing in his work. Ashuach has led in both product and business innovation with his online platform, UCODO, that demonstrated changing customer interactions using additive manufacturing. Lionel Dean's work at FutureFactories.com demonstrates the innovations possible with the technology, and Michael Eden, who designs contemporary interpretations of historical pieces realised with additive manufacturing, illustrates it is possible to design products that have community value, as both Dean and Eden have pieces in museums in Europe and the United States.

For individual consumers, additive manufacturing is well placed to provide the means to have an invested product created for them, or by them. One of the key features of additive manufacturing is its ability to enable the fabrication of one-off items without the constraints of tooling costs that hamper conventional manufacturing. Dean demonstrated how to achieve this with his use of animations controlled by the customer to achieve a design to their liking.

Nervous System takes this to the extreme, as this type of interaction is made possible directly on their website. This enables the customer to literally push and pull an object to shape before requesting a 3D print. For the professional product designer, providing the facility for a bespoke invested object to be created involves a mind-shift in practice from designing a single resolved product offered for sale to an unknown buyer, to designing a body of work on a single design concept that is validated within set parameters. The customer would have the freedom to specify within those limits a product bespoke to their taste. The product may be based on personal data – for example, for 3D printed insoles or hearing aids – or may be based on selections controlled by online interactive sites.

Examples of invested objects include adding the ability to personalise text to a 3D printed product, such as in light shades produced by Materialise and the ability to add bespoke ergonomic features, such as for manufacturing jigs. Creating an emotional or financial attachment to an object, or creating a feature that adds a functional personalisation, has the potential to increase the life of that object and, therefore, maximise the embodied energy invested in its fabrication. Tom Dixon described this approach in relation to his work "Fresh Fat," which is, arguably, fabricated using a form of additive manufacturing, though only loosely by definition, as it is built from a free-form extrusion controlled by Dixon, rather than a layer-by-layer process and automation. However, as gel 3D printing and free-form robotic arm printing become more prevalent, Dixon's Fresh Fat appears more connected. For example, the Fresh Fat chairs were made from 30 kg of extruded polyethylene terephthalate glycol (PETG) co-polyester, which is not exactly best practice in sustainable design by any definition. However, Dixon argued that because of his reputation, the chair would be kept intact for many years and, therefore, was a worthwhile investment of materials that justified his description of it as sustainable by design.

Another approach to sustainability is illustrated by Dixon's other high-profile sustainable project. When Dixon took over leadership of the furniture company Artek, he advertised to buy back the original mass-produced stools the company had made in an initiative called *Artek Second Cycle*. This simple but elegant stool was bought mostly by institutions such as schools, but also by some individuals. He offered to buy the stools back on condition that they came with a record of where they had been. Their individual history was then added to a plaque attached to the stool. By adding the story of the stool, Dixon was demonstrating how to create an invested item for the customer, even from an object that was not individually connected to them. This added social/financial value that would ensure the product was retained. Using this approach, but viewed through the additive manufacturing lens, the facility for an individual to be able to 3D print their own copy of an object, using a desktop 3D printer, could arguably add a similar value. Even if an element of a product could be fabricated this way, the customer could have a stronger attachment than to a product bought from a shop. Some of these ideas are already being explored through the sharing of 3D printable files as non-fungible tokens (NFTs), which permanently store the history of ownership of the item in a digital ledger on a blockchain.

Open-source file sharing can also accommodate this intrinsic value, with people able to freely download and modify some designs through appropriate Creative Commons licenses, sharing their modifications as remixes, and building libraries of products as an online community. The most well-known example of this is #3DBenchy, a model of a tugboat mentioned in the "accuracy" paragraph of the fixtures section in Chapter 3 that began as a popular test piece for 3D printers but now has a life of its own with countless variations and add-on parts.

For larger companies, 3D printing offers the ability to produce an element based on customisation, including customisable dashboards and other details. In this way, it may be possible to add to the sustainability credentials of a product. However, as the parts are essentially plug-ins, it is unlikely that it will create the same level of emotional investment for the customer that a more hands-on, or in-depth, connection would. BMW tried this approach, and despite gaining considerable attention for the company, its viability is challenging at this time.

Supported by practical developments in associated digital technologies, such as 3D scanning, data generation, and 3D augmented reality, ideas and understanding around the potential of the technology to enact change beyond the technical has begun to emerge. It has finally shifted in its perception from a prototyping tool into a form of industrial practice. However, there is frequently a disconnect in the literature between its practical use and its ability to disrupt conventional practice. Much of the focus has been on building its commercial capabilities as an additional technology to complement or replace existing manufacturing practices, without rethinking how it could help shape developments in society more broadly in response to current aspirations for societies across the globe.

BEYOND COMMERCIAL PRACTICE

Shaping the fourth industrial revolution to ensure that it is empowering and human-centered, rather than diverse and dehumanizing, is not a task for any single stakeholder or sector or for any region, industry, or culture. The fundamental and global nature of this revolution means it will affect and be influenced by all countries, economies, sectors, and people.

[7]

Additive manufacturing, taken to the extreme, has the potential to provide the foundation for more egalitarian production practices. It has the potential to enable a more equitable distribution of control in the manufacture and distribution of products. Yet, its revolutionary potential is overshadowed by the commercial ambitions of individual companies involved in its development. Governments are not yet shifting their ideologies towards intergenerational responsibility, and manufacturing is still viewed as a short-term fragmented commercial activity rather than intricately connected to the sustainability of life on the planet. Beyond environmental sustainability, disproportionate control of the means of production and subsequent wealth appear to be seen as necessary constructs of the free market. Additive manufacturing, taken in its purest form, could help rebalance the world, if there was a collective will to embed it into smart cities; and change the relationship between products and consumers, as it could enable a more socially responsible approach to the role of the individual and communities in shaping consumerism.

These are idealistic notions, where additive manufacturing allows individuals and communities to break free of the dominant production and consumption paradigm to have more control of products for their own use. Yet, there is evidence that changes in thinking about products and their relationship to individual needs and societal aspirations is emerging in this space. Society 5.0, which started in Japan as a government initiative to frame a fairer, more sustainable, more comfortable society enabled by digital technology, is a prime example. The Society 5.0 philosophy highlights the connection between people and things and merges the real and the cyber. These ideas are seen as essential by the Business Federation in Japan to help resolve issues in society, create better lives for its people, and sustain healthy economic growth. However, the success of the approach also depends on gaining commitment across different levels of society and could, therefore, remain a utopian vision.

Although the challenge is to build commitment to Society 5.0 for a digital era, additive manufacturing is yet to be seen as central to this ideal. However, the personalisation and distributed manufacturing capabilities of additive manufacturing align to this approach. With the growing global population and the subsequent demands on natural resources, additive manufacturing could sit at the centre of the strategies of Society 5.0 to maximise materials and personalise output. The environmental impacts of methods of production, patterns of distribution, and consumption across communities became increasingly visible as lifecycle assessment practices evolved, thanks in part to digital connectivity and data collection. The pollution and waste created over the last 100 years have accumulated enough to raise alarm. The arguments for maximising opportunities to help dematerialise products, and product service systems, through digitisation have grown over existing commercial interests. This has largely been driven by shareholders.

Since the turn of the century, manufacturing has been edging towards designing people out of production systems, with increased automation taking the means of production far beyond the power of individual workers. According to Nigel Cameron [8]: "We're at the outset of a great debate. At one level, it's simple. It's about whether we need to worry that robots will take our jobs, or whether we don't." The emergence of Society 5.0 suggests that, at least in some quarters, the unexamined adoption of automation by independent commercial interests could be damaging to communities. Arguably, for the benefit of societies, companies should be monitored for environmental protection purposes as well as elements relating to social engineering, and evidence of sustainable product production systems for the benefit of society more broadly.

> Technological design choices are, in effect, social choices with significant political ramifications and distinct ethical valances. They make clear that in designing and developing technical objects and systems, we are doing more than merely creating new tools to be put at our disposal; we are affecting the social fabric, and indeed our own individual agency, in morally relevant ways.
>
> [9]

Industry 4.0 was introduced in German manufacturing to maximise the output of a production line through monitoring and responsive quality control systems. As sensors have become ubiquitous, data collection and analysis has informed the spread of Industry 4.0 as an approach beyond the confines of manufacturing itself. Product lifecycle, market demands, and distribution patterns can now be used to shape supply. Even product performance and maintenance can be monitored, and problems mitigated prior to critical events. Additive manufacturing is central to the suite of connected digital tools available for use within current production systems. The very ability to closely monitor and determine details based on an Industry 4.0 approach lends itself to the underlying principle of additive manufacturing, which is that a part can be individually fabricated to maximise the material being used for a bespoke need. In addition, the ability to share files, and build co-operatively, means that the technology could support more equitable control over the means of production in society.

The challenge is whether there is the collective will and political determination to use these abilities for the benefit of society rather than leave it in the hands of commercial enterprise. The emergence of Society 5.0 suggests that, at least in Japan, a political will is starting to respond. The ontology of the Product Design discipline demonstrates a change towards supporting more sustainable, socially responsible products and systems. Digital technology is not as central to the Product Design discipline as it could be, yet the discipline's critical role in manufacturing means that developments in the digital era should be integral to education and future practice.

For the product design profession, additive manufacturing creates two main problems. On the one hand, the large body of knowledge developed within the discipline over the last two centuries does not correspond well with the body of knowledge needed to work with additive manufacturing technology. The skills and understanding needed to work with conventional manufacturing practice and technologies do not build naturally on those of additive manufacturing, creating a disconnect, or discontinuity, in the professional development of the discipline. This results in a lack of leadership within the discipline on its use.

Resistance, as discussed in working with existing practice in Strategy 1, can be present in the product design discipline itself. Senior designers may feel as threatened as production practitioners do when a radically new way of working emerges that could undermine their authority. On the other hand, additive manufacturing has not appeared as a single technology but rather a whole mess of complicated, and barely connected, technologies that all require subject expertise to work with effectively and at a professional level. The product designer has to understand designing for the different constraints and opportunities of each technology they come in contact with, and they also have to understand their different impacts on business practice and understand their role as disruptive technologies.

> There is little disagreement that many disciplines are witnessing an "ethical" turn. The raised awareness of environmental degradation and climate change, inequalities in wealth distribution and persisting conflict around the world have caused individuals who are practitioners in different disciplines to question their ethical obligations, not merely to those close to them, but to global "Others" whom they cannot fully define.
>
> [10]

There is, therefore, a strong argument for manufacturing to fundamentally change in the twenty-first century to benefit society and reduce its impact on the environment in a world facing the current environmental crisis. Additive manufacturing could be central to changing practice and the organisation of production and distribution, the use of resources, and the codesign of products for individuals over wasteful mass-production systems. The technology could be part of planning for the future in an increasingly digital era by improving inclusivity through the means for greater economic independence, where individuals and communities are better able to meet their own product needs.

During the last decade, additive manufacturing has evolved from prototyping to end-use manufacturing. Production-scale machinery has emerged, and the range of technologies defined as additive manufacturing has expanded. However, whilst standards have been established, and research into quality control systems has increased, including responsive systems within the actual build chamber of metal powder-based printing itself, the aim appears to be to replicate the quality and reliability of traditional manufacturing techniques over exploring new design fields. Investment in industry remains predominantly on machinery and new materials, such as composites, over changing business practice for sustainability reasons.

Changing the established manufacturing systems involves changing priorities in market and supply practices, both ethical and moral. The emergence of design-focused groups such as the Designer's Accord and the Living Principles, which aim to "foster creative action for collective good," combined with the increasingly overt effects of climate change, suggest that the conditions may be changing for a rethink of standard practice. This is supported by the global megatrends of moving towards sustainability ("more from less" and "planetary pushback") and personalisation, identified in 2015 by Stefan Hajkowicz [11]. As members of the digital Generation Z move into positions of power, and

Millennials begin to outnumber those in the Baby Boomer generation, priorities may change, and visions of a more sustainable future enabled by digital technology might not seem so far-fetched.

SUSTAINABILITY IMPERATIVE

What is best practice in the design of consumer products? What exemplifies cutting-edge contemporary ideals and practices in manufacturing? Arguably for both, it is an outcome that demonstrates two things. First, product design and production systems are informed by current knowledge. Second, the work reflects the aspirations of a society at that point in time. A product should, therefore, be judged a success or failure in relation to its context rather than in isolation.

Over the last two decades, there has been rising acceptance of the growing sustainability imperative driven by climate change concerns. In 2007, the Intergovernmental Panel on Climate Change (IPCC) produced the Synthesis Report during the fourth assessment report cycle (AR4) and warned that "warming of the climate system is unequivocal, as is now evident from observations of increases in global average air and ocean temperatures, widespread melting of snow and ice and rising global average sea levels."

Whilst additive manufacturing, as a technology alone, is not going to impact the planet, what a paradigm shift towards additive manufacturing can represent could, in theory, change entrenched patterns of production, consumption, distribution, and disposal. It may be an optimistic view, but it is one that needs articulating, as it could contribute to new ways of consuming resources in the future. It is also topical now because of the generations coming through, who are more interested in a dematerialised economy and appear more open to radical changes in behaviour to protect the planet. The current context, therefore, is that alarm bells have been sounded on the environmental impact of continuing with production and consumption as it stands, and the next generation is more willing to consider alternative ways of working than the older generations. Add to this the digital revolutions that have occurred, and the convergence of digital technologies over the last decade, and more radical possibilities are emerging.

With regards to sustainability, researcher Fuad-Luke [12] provided an interesting review of design practice over the last century through the lens of current sustainability knowledge and philosophy. He evaluated the work being undertaken at a particular point in time, not in relation to the attitudes and knowledge that was prevailing at that time, but in relation to current knowledge and ideas. By projecting this approach forward, it is interesting to consider what a similar sustainability evaluation of the industrial practices and design outcomes happening now look like from a vantage point 50 years into the future. As disruptive digital technologies converge, the hope is that industry, communities, and governments collaborate to alter patterns of production and consumption that are damaging to the environment and promote inequality in society. Additive manufacturing could arguably play a critical role in changing the current manufacturing paradigm to better reflect sustainability aspirations for society, and to address meeting the sustainability imperatives needed for succeeding generations.

A challenge in answering these questions lies in whether it is possible to critically engage with the digital developments of the last 20 years enough to be able to effectively realise the potential of those technologies within the legacy of existing industrial practices. Humanity would need to step outside the complicated economic web it has created after the industrial revolution, and the rise of mass manufacturing and assembly line production, to evaluate its position in a digital era objectively. This would require balancing production systems and patterns of consumption against the impact of mass-manufacturing practices, and a throwaway society, on the environment. The broader rise of industrialisation and its impact on urbanisation and the structures of societies around the world have been thoroughly analysed, and alarms about the environment and sociocultural sustainability have been raised by thought leaders at this time, such as environmentalist and politician Al Gore [13]. It is interesting to consider whether industrialisation would have progressed in the same way through the twentieth century if societies had known the extent of the environmental and social problems that would stem from this paradigm, and if humanity would have been able to redirect its efforts towards more sustainable patterns had that been the case.

The last 20 years have seen a digital revolution brought about by developments in technology across broad areas. These include communication, data collection and analysis, digital fabrication, and computing power. These developments may initially have appeared disparate and purely technological, but they are increasingly converging and creating visible change to societies. Examples include the evolution of the Internet from a straightforward communication tool into a powerhouse for reimagining society, the development of sensor technology so low-cost that new cars typically have thousands of sensors in the engine feeding back information to the user and to the manufacturer. They also stretch to the computing power for the masses through smartphones and wearable technology.

Digital technologies are now integral to interaction. The complexities of a digitally enabled society, and the potential of technology-enabled communication to influence attitudes and behaviours, are clearer. There is also a growing awareness of their influence on society, from education to politics. The conditions for a conscious reset of the attitudes and behaviours that create production practices, patterns of consumption, and working conditions for the growing sustainability imperative are emerging. Arguably, it is time for a rethink in response to the opportunities and challenges of a digital era.

One of the indicators of sustainability attitudes and aspirations at this point in time is the content of high-profile exhibitions, and here additive manufacturing is predominantly featured. The 2013 *The Future Is Here* exhibition at the Design Museum in London illustrated an essentially optimistic view of the role of technology in a future society at that time, including the heralding of mass customisation and desktop 3D printing as the start of a democratisation in making. By 2018, the London exhibition *The Future Starts Here* at the Victoria and Albert Museum was far more dystopian. The role of technology appeared ambiguous with respect to the aspirations for society, described by the tenor of the displays. Personalised 3D printing was again featured, but this time the presentation was more subdued in its ambitions, addressing bespoke assistive technology needs rather than laying claim to a radical overhaul in the control of product manufacturing.

The hype of digital fabrication appeared to be fading, replaced by more commercial applications of the technology. However, over the last two decades sustainability researchers argued that there needed to be more of a focus on dematerialising society, and additive manufacturing has the potential to be central to this approach. Yet, 3D printing industry-led events and research have suggested little appetite. The major push in industry over this time has been towards a mass-production equivalency for the technology. Efforts have focused on increasing machine throughput, and the automation of processes, from topology optimisation to post-processing, even though this is not the unique selling point of the technology. A high-profile example of this has been the Adidas Futurecraft shoes, where the plan was to use Continuous Liquid Interface Production (CLIP) technology to produce shoes en masse using a non-recyclable material. In this case, 3D printing was essentially being used in the same way as a mass-production technology, with a marketing spin of customisation.

A push towards mass-production practices for 3D printing hijacks the potential of digital fabrication tools to help reshape society towards a less wasteful, less expensive model. Driven by commerciality, industry tends to present just one approach, one solution to meeting material needs, despite there being a very different path available if society could turn towards dematerialisation. As a concept, Industry 4.0, or Industry 5.0, or Eco-efficiency x Factor X, can utilise digitally enabled manufacturing, and lifecycle tracking, to redirect patterns of production, consumption, and resource management to better meet society's environmental and sociocultural sustainability needs. Rather than for efficiency in industrial practice to maximise profit, Industry 5.0 could provide a very different blueprint for a leaner, circular economy as an ethical foundation for planning the health and well-being of future generations. To exploit these opportunities, a rethink of the relationship between manufacturing and the consumer is needed, retraining humanity's consumption patterns enabled by the tools available in a digital era.

3D PRINTING FUTURE SUSTAINABILITY

After World War II, US President Eisenhower urged "Buy, buy anything" to boost the economy [14]. The need to stimulate consumerism led to products having built-in obsolescence to increase sales. An example of this was the introduction, each year, of new stylistic versions of a make of car. This approach, combined with an increase in disposable products, such as disposable packaging for takeaway meals, led to the rise of a "throwaway" society. Vance Packard, in his early and highly influential book *The Waste Makers* [15], criticised these evolving patterns of behaviour and called for more responsible design.

Imagine if the environmental impact of the industrial revolution known now was known at the start of industrialisation. Would manufacturing and the organisation of labour, and patterns of consumption and disposal, have been designed differently? Hundreds of years later, this is the start of the digital convergence revolution. So, projecting forward, is it possible to anticipate the impact of these seismic shifts on practice and take steps to shape the future for the benefit of the planet, and all its occupants? What would that look like, and what would it mean for developments now to attain that desired future? Environmental and social sustainability are increasing concerns, as are concerns over the impact of the digital revolution. Environmental, health, and global political pressures impact the way manufactured goods are made, distributed, and consumed. Does 3D printing have the potential to contribute to future proofing forms of manufacturing and consumption in the face of these issues? And if so, how?

If economies went "all in" with additive manufacturing, it would create a radical change to production systems. An additive manufacturing-based paradigm could, theoretically, reframe patterns of production and consumption in the long term. In the short term, it would help sustain manufacturing when material supply and downstream demand are unstable. Product designers who are actively designing for the circular economy, and extended producer responsibility, are responding to global megatrends that have influenced the impact of 3D printing for sustainability and could potentially impact the role and output of designers in the future, were that paradigm shift towards an additive manufacturing approach to production and consumption to occur.

This idea provides a critical view of existing production systems and attitudes and is, therefore, potentially provocative. It draws a line under the industrial revolution and mass production. It further outlines the reasons for a radical rethink of design, supply, consumption, and product disposal. However, if there is a suspension of disbelief by the discipline to consider the digital convergence that is currently building, then it may provide some insight into ways forward if the environmental imperatives are to be seriously addressed.

Socially responsible design strategies based on a holistic approach to production, use, and resource recovery consider the effect of design decisions on all those affected by the way the product is resourced, produced, distributed, used, and disposed. The social impacts on those working in manufacturing, as well as the impacts on those affected by the harvesting of materials, form part of this holistic accountability. As far back as 1918, Pigou, an economist based in Cambridge, United Kingdom, proposed that lifecycle assessment should include a social impact assessment element. He argued that the costs to society, associated with welfare economics, should be embedded in the system, and made visible. Over 80 years later, the Institute of Applied Ecology built on this in a triple product line analysis proposal that included environmental, societal, and economic impacts. John Elkington, in his book *Cannibals with Forks*, described this as triple bottom-line accounting [16].

Papanek's seminal work from 1972, *Design for the Real World: Human Ecology and Social Change*, in which he questioned the social responsibilities of designers in a global production environment, positioned these issues as central to product design, and it is still in print today [17]. Papanek proposed that, in general, societies had more than enough products, and product designers should focus on solving humanitarian problems. The impact of his ideas is illustrated in humanitarian projects, such as those described in *Design Like You Give a Damn* [18]. Emerging digital technologies provide an opportunity to radically rethink practice for a more sustainable future, including environmental conservation and community good. The problems Papanek focused on, and tools he had available, may have evolved, but the social responsibility he advocated for is just as critical, and arguably more urgent. Business as usual is just not working for the planet.

NOT BUSINESS AS USUAL

Additive manufacturing encompasses a range of digital fabrication technologies that build objects in a layer-by-layer process without the need to preform tools, or to use traditional tooling [19]. This allows for a significant shift in what can be produced, by whom, and where. This facility enables radically different business models that are only starting to emerge. From a sustainability point of view, this facility allows for print-on-demand capabilities that, in theory, could create a paradigm shift in how, why, when, and where products are produced. This aligns with the drive for the dematerialisation of the economy as a leading sustainability strategy. Ryan [3] argued his six sustainability principles were the basis for creating a viable basis for an economy that would change practice. The principles were:

- Valuing prevention.
- Preserving and restoring "natural capital."
- Lifecycle thinking (closing system cycles).
- Increasing "eco-efficiency" by "factor x."
- Decarbonising and dematerialising the economy.
- Focussing on design – of products and product service.

Additive manufacturing capabilities for on-demand production fits these principles very well. However, little connection has been made between the radical sustainability capability possible in 3D printing and ongoing industrial development. This is because the technology has been largely developed by engineers

8.5
At-home, desktop
3D printer fridge
vent repair (Loy).

and implemented in fields where sustainability is low on the list of priorities (e.g., medical, aerospace), based on predominantly commercial needs within established organisational practice. As the investment required to develop new technologies is considerable, this is not surprising. However, as the sustainability imperative grows for governments, societies, and the environment, the radical rethink in operations, possible with additive manufacturing at its core, may be of greater interest.

In the future, the alternative to mass production is high-value, bespoke, invested products, made collaboratively in a distributed manufacturing model, supported by digital connectivity and file sharing, which may be viewed differently. The COVID-19 pandemic certainly demonstrated that operations could change quickly and in unexpected ways when there is sufficient pressure to do so.

Sustainability design with additive manufacturing models a shift from a throwaway society to a society in which each product is valued, produced only when required, and can be repaired (illustrated in Figure 8.5), updated, or fed back into the process to reduce landfill. To sustain the economy, this would necessitate a change in where the value of a product is emphasised. Essentially, it would entail moving the value from the material to the intrinsic value created through collaboration and investment of time and skill required to produce it. This creates a society of consumers who work with the maker to design the products, and prosumers who both produce and consume their own products [20]. However, it requires a reframing of consumerism which would be possible only with a seismic shift in priorities driven by external forces, such as disruptions to supply chains brought on by the COVID-19 pandemic. Climate change may also be a key driver of this reframing and shift as the negative effects of by products from industrial processes are now better understood, and the effect of unsustainable practices on pollution and waste dumped in oceans is becoming more out of control. The following statement is indicative:

[A total of] 8300 million metric tons (Mt) of virgin plastics have been produced to date. As of 2015, approximately 6300 Mt of plastic waste had been generated, around 9% of which had been recycled, 12% was incinerated, and 79% was accumulated in landfills or the natural environment. If current production and waste management trends continue, roughly 12,000 Mt of plastic waste will be in landfills or in the natural environment by 2050.

[21]

If an alternative manufacturing and consumption paradigm exists which maximised the use of raw materials with a set of technologies such as those in additive manufacturing at its core, what would it look like? If this approach was embraced by societies, what could happen, and working backwards, what would need to happen in the meantime for this to succeed?

The basis for mass production was arguably ideologically flawed as it relied on producing quantities of generic products for unknown individuals, often with little evidence of there being any demand. Therefore, there was little or no relationship between customers and products. The selection of details between products, such as different colours or finishes, were limited and predetermined. Mass-produced products, even those customisable components, were not personalised. In other words, they were manufactured without specific information pertinent to the user but, rather, were fully resolved by the designer, albeit with several options.

The legacy that is weighing over the digital revolution is consumerism, which emerged from the drive for economic growth following World War II. At this time, the pressure to consume more to promote wealth, and for consumer products to have built-in obsolescence to support the economy, was framed as a positive approach by governments. The paradigm shift that needs to occur for a radical rethink of consumer behaviour is to change the view that products are disposable, to one where products are seen as "continuous."

ADDITIVE MANUFACTURING CONTINUOUS PRODUCTS

If products could be viewed as "continuous products" rather than as objects with a limited useful life, and there was more personal and collective responsibility, from manufacture, maintenance, and adaptation of products, it would create a paradigm shift in production and consumption. This approach could be incentivised by governments but would require a very different ethos, and a break between mass producers and policy makers that may be possible only if an external impact, such as the effects of climate change, provides the driver.

Additive manufacturing is considered to be promising for sustainable production because the additive and digital nature provides opportunities to save resources. This additive and digital nature enables, for instance, on demand production of spare parts for repair … or avoids material losses when compared to subtractive technologies such as milling.

[22]

If people were restricted in what they could buy and were no longer able to buy and discard mass-produced products without consequence, then products would be valued and maintained. Additive manufacturing could enable a greater connection with products, especially as distributed manufacturing machinery improves from low-cost desktops towards shared and community-based machines. Products could be designed to be adapted, and product service systems evolved to support those unable to make those changes themselves.

In addition, additive manufacturing has evolved as part of the maker movement, which has a significant following in the United States and the United Kingdom in particular. Neil Gershenfeld from MIT established the first FabLabs in 2001 in Boston to provide workshops for the public to use digital fabrication technologies, including 3D printing. He has also developed his ideas with colleagues on designing reality in a digital era since then [23]. At the time of writing, it was estimated there were over 2000 FabLabs worldwide, registered with the Fab Foundation, although it is likely there are many more. File sharing expanded rapidly since 2010, when the first 3D printing public online service providers were launched [24]. These facilities model a community practice engaged in developing a continuous products philosophy, as does the work in the United Kingdom and Europe on the right to repair consumer goods without voiding their warranty. Although, at this time, the approach is happening at the fringes of consumer society, with pressure from the growing sustainability imperative, it could be a model for extended practice in the future. For product designers, the drive aligns with the development of sustainability and ethical principles as the foundation for designer-focussed principles such as *The Designer's Accord* and *The Living Principles*, given their mission statement of "creative action for collective good" [25].

Additive manufacturing maximises the use of materials, as value-added bespoke products are created on demand. This contrasts with the wasteful practice of mass production, where products may not be sold and go directly to landfill without ever being used. In addition to environmental sustainability drivers, arguably additive manufacturing provides societies with the means to redistribute the means of production. This could enable individuals to produce their own goods as needed and reduce their outgoing costs. Where automation designs out people from manufacturing processes, there is an argument for individuals to be able to operate more independently from the conventional consumer economy. In addition to supporting the dematerialisation of the economy, additive manufacturing could have impacts on the organisation of society, especially as part of smart city regeneration with individuals creating their own income from home by leveraging digital technologies. Centralised manufacturing could predominantly be replaced by community-based distributed manufacturing, and a product service system approach.

This approach to dematerialisation of the economy is not as outlandish as it may have seemed two decades ago, as recent shifts in consumer demands arguably laid the foundation for a shift in product manufacturing. The rise in streaming services, usurping production of CDs and DVDs, for example, and the introduction of connected products in people's homes and transportation, laid the groundwork for the dematerialised economic model. COVID-19 made changes too, with online shopping reducing the need for physical premises for retailers. Significant changes would be needed to reframe societal priorities and redirect policy towards building a different manufacturing paradigm. Consumerism would need to be revised, and products would need to be valued, and designed, differently. A community of practice would need to be built with very different ways of interaction, based on changed aspirations for society to meet the sustainability imperative alongside the means enabled by the digital era.

In *The New Tin Ear* [26], Aldersey-Williams discussed additive manufacturing as a way for communities to return to more localised production, describing mass production as a misstep in human evolution because of its impact on the environment and the organisation of societies. Meeting the growing sustainability imperative does require change in behaviours, and digital technology is creating disruptions in business practice. However unlikely it may seem, in the face of the hold that large corporations have, it may be that with external pressures, additive manufacturing can emerge as the basis for the transition to more sustainable, user-centred, localised manufacturing product service systems.

The dismantling of Eisenhower's throwaway society needs to be based on practical strategies and new policies for it to be effective. The collective impact of product designers, not only on the environment but on social engineering as well, needs to be considered. The initial focus for sustainability was on the environmental effects of material sourcing techniques, and then on pollutants expelled during processing, then on waste, material reclamation, and disposal. However, the impact of current industrial practices and systems, as well as patterns of production and consumption on shaping the economy, and subsequently societies, has been under increased scrutiny in recent years.

As Martin Charter, Director of the Centre for Sustainable Design in the UK, suggests, the term "Circular Economy" (CE) implies product service systems need to be rethought and sustainability integrated to practice. He argues for a society where "product life extension through repair, refurbishment and remanufacturing is the prevailing social and economic model" [27]. Digital convergence, the growing urgency of the sustainability imperative, and changing behaviours and priorities within society can support a radical new way of thinking and operating, if the collective will has finally arrived.

REFERENCES

1. Hawken P. *Drawdown: The Most Comprehensive Plan Ever Proposed to Roll Back Global Warming.* London: Penguin; 2017.
2. Ryan C. *Digital Eco-Sense: Sustainability and ICT – A New Terrain for Innovation.* Melbourne: Lab 3000; 2004.

3. Benyus J. *Biomimicry*. London: Harper Perennial; 2002.

4. Rael R, San Fratello V. *Printing Architecture: Innovative Recipes for 3D Printing*. Princeton: Princeton Architectural Press; 2002.

5. Walker S. *Sustainable by Design*. London: Earthscan; 2006.

6. Bohm F. *KGID: Konstantin Grcic Industrial Design*. London: Phaidon Press; 2005.

7. Schwab K. *The Fourth Industrial Revolution*. New York: Crown Business; 2017.

8. Cameron N. *Will Robots Take Your Job?* Cambridge, MA: Polity; 2017.

9. Morrison L. Situating Moral Agency: How Postphenomenology Can Benefit Engineering Ethics. *Science and Engineering Ethics* 2019, https://doi.org/10.1007/s11948-019-00163-7.

10. Loo S. Design-*ing* Ethics: The Good, the Bad, and the Performative. In Felton E, Zelenko O, Vaughan S, editors. *Design and Ethics: Reflections on Practice*. London: Routledge; 2012. pp. 10–19.

11. Hajkowicz S. *Global Megatrends: Seven Patterns of Change Shaping Our Future*. Melbourne: CSIRO Publishing; 2015.

12. Fuad-Luke A. *Design Activism: Beautiful Strangeness for a Sustainable World*. London: Routledge; 2009.

13. Gore A. *The Future*. London: WH Allen; 2013.

14. Sparke P. *An Introduction to Design and Culture: 1900 to the Present*. 3rd ed. London: Routledge; 2013.

15. Packard V. *The Hidden Persuaders*. Reissue ed. New York City: Ig Publishing; 2007.

16. Elkington J. *Cannibals with Forks*. New ed. Knoxville: Capstone; 1999.

17. Papanek N. *Design for the Real World, Human Ecology and Social Change*. London: Thames and Hudson. 1999.

18. Architecture for Humanity. *Design Like You Give a Damn*. London: Thames and Hudson; 2006.

19. Diegel O, Nordin A, Motte D. *A Practical Guide to Design for Additive Manufacturing*. London: Springer; 2019.

20. Aspelund K. *The Design Process*. 2nd ed. London: Fairchild Books; 2010.

21. Geyer R, Jambeck J, Law K. Production, Use and Fate of All Plastics Ever Made. *Sciences Advances* 2017;Jul 19;3(7):e1700781.

22. Sauerwein M, Doubrovski E, Balkenende R, Bakker C. Exploring the Potential of Additive Manufacturing for Product Design in a Circular Economy. *Journal of Cleaner Production* 2019;226:1138c1149.2019.

23. Gershenfeld N, Gershenfeld A, Cutcher-Gershenfeld J. *Designing Reality: How to Survive and Thrive in the Third Digital Revolution*. New York: Basic Books; 2017.

24. Stacey M. The FAB LAB Network: A Global Platform for Digital Invention, Education and Entrepreneurship. *Innovations: Technology, Governance, Globalization* 2014;9:221–237.

25. Loy J. The Changing Role of the Designer, "Breakthrough Innovation" and the Technological Tools for Change. In: Anderson L, Jackson S, editors. *The New Design Nexus: ICT, Changing Demographics and Sustainability*. Melbourne: Lab 3000; 2006. pp. 54–57.

26. Aldersey-Williams H. *The New Tin Ear: Manufacturing, Materials and the Rise of the User-Maker*. London: RSA Design Projects; 2011.

27. Charter M, Keiller S. *Repair Cafes: Circular and Social Innovation, Designing for the Circular Economy*. London: Routledge; 2018.

9 Making the future/remaking product design

In this final script, the CEO, production manager, and in-house designer from Chapter 2 are meeting again with Meeks, the designer from LND Consulting. The topic is adapting their business model to the changing industrial and business landscape, which is not an easy idea to discuss. Meeks walks in whilst they are mid-argument.

Production manager: I'm just saying, it's not as easy as you think.

In-house designer: I don't see we have an option. If our supply chain is interrupted again, we'll have more of a problem if we're not prepared.

Production manager: I don't see our competitors investing in more technology at the very time things are tight. Most people are laying off staff.

In-house designer: Meaning?

CEO: Come on, you two, and yes, your jobs are safe, but we do need to continue to adapt. It feels like things are changing quickly in the global landscape.

In-house designer: With more investment, we can shift to fully agile production practices.

CEO: I think you're getting ahead of yourself, but we could look at the possibilities. (*Turns to Meeks.*) Meeks, thanks for coming. What do you think?

Meeks: I'm thinking there are interesting possibilities if you consider shifting the business to being agile and adaptive. It would be a different business model from what you're used to, and we'd have to build greater functionality into the online platforms, plus redesign products for 3D printing.

Production manager: I'd like more digital inventory; that's worked well for us.

In-house designer: It goes beyond more components though.

Meeks: True. In fact, with the ability to design to an individual scenario, we're seeing more companies develop the product service system side of their business using the technology.

CEO: That doesn't sound like something manufacturers like us do.

Meeks: It is moving the focus from making products and then trying to sell them, to a more collaborative relationship with your customers, but you've done well in the personalisation space.

DOI: 10.4324/9781003122203-10   **266** ☐

In-house designer: Very well.

CEO: Though it's not been without its challenges.

Meeks: I'm just saying, as your knowledge and experience in using 3D printing, as well as other digital technologies in the workflow grows, you can begin to add value through shifting the focus from the product to the customer.

Production manager: And now we're back at complications again.

Meeks: It could have other benefits as well.

CEO: Such as?

Meeks: Legislation such as extended producer responsibility means customers will be looking at maintenance and repair differently in the future. We're already seeing online communities redesign parts for 3D printing and share them so they can repair their own products. Thinking bigger, what happens when this becomes the norm, when providing this level of support to customers becomes legislated?

In-house designer: Good point, and maybe we'd have customers for life instead of just a transaction.

CEO: Sounds like a logistical and legal quagmire to me, but let's talk it through.

Meeks: I know things have changed over the last few years, but you've adapted. I wish I could tell you it's okay to consolidate your position and dig in, but …

CEO: Alright, let's do this.

Production manager: I'm going to need more coffee.

PRODUCT DESIGN DIGITISATION

If recent years have taught us anything, it is the unpredictability of the way the world will be in the future. This is an added complication for product designers, as the product design lens is not shaped solely by technical opportunities but by user experiences, aspirations for society, and environmental protection. Whilst creative possibilities are provided by evolving, and converging, digital technologies, designers are not only concerned with the object in isolation. This inspires the question: what is different for product designers working with additive manufacturing in the current and emerging context when the pace of digital developments is so rapid? What opportunities and challenges await the discipline?

> There is a clear consensus that the future now emerging will be extremely different from anything we have ever known in the past. It is a difference not of degree but of kind.
>
> [1]

For product designers operating between the inception of the profession in the 1770s and the end of the twentieth century, the scope of the role within manufacturing was clearly defined. Initially driven by the need to provide commercial outcomes to a widening customer base and suitability for mass production, the focus was on designing and optimising products that could be reliably made in factories, built on division of labour principles that could also respond to market trends in art and décor [2]. Beyond these commercial needs, there was little imperative to consider the impact of design decisions, and the focus was on the practicalities of manufacturing and product sales.

Increasingly, however, the role of designers involves looking beyond the technical aspects of manufacturing to the user experience, and to the market, as well as the impact on society and the environment. Fundamentally, it provides a practical response to the challenge of combining the arts, humanities, and science to solve problems, without aligning solely to any other individual, or existing discipline. Therefore, in the last 30 years, the scope and definition of design practice has broadened further. Product design education has increasingly addressed the development of product service systems over products, of building sustainability ecosystems, supporting co-creation, and investigating the role of design in shaping society [3]. In the last 15 years, there has also been a significant change in the relationship of design and business, with high-profile design companies such as IDEO reframing design as "design thinking" [4], business strategy, and innovation. In addition, futurists have begun commandeering the design space to explore speculative work in environmental, political, and social activism [5]. This has been alongside developing practice in co-design, values-based design, and, most recently, speculative design [6].

As the discipline has broadened, it has shifted in its purpose, and its perspective, and has arguably become fractured. Some design programs have become more industrial and aligned with engineering, whilst others have become more aligned to the creative industries. Yet, others have become focused on social and environmental problem solving, with less emphasis on physical products at all. In this sense, there is no longer a consensus of what a product designer does compared to the previous two centuries, or what skills they should have. Business schools now teach design thinking [7], and engineering schools teach product development [8] as well as product engineering.

University product design program leaders are faced with the dilemma of where the program should be positioned. Some have embraced the design for need and sustainability principles championed by Papanek in the 1970s [9], and others focus on entrepreneurial skills over industrial ones. It is not yet clear where the discipline will coalesce in the twenty-first century, or even if it will, or whether it will become further fractured to the point where it is simply distributed between other disciplines. Certainly, the push to teach design thinking and product development to all students in universities (e.g., University of Technology Sydney, Australia: Transdisciplinary Design) as the basis for innovation in entrepreneurial studies could lead to a dilution of what is product design, or a reduction of the discipline to being viewed as a manufacturing feeder.

In recent years, product designers have also shifted their collaborations from mechanical to electrical engineering, and even to information technology (IT). This has further complicated program organisation as connected products came to the fore, and terms like human-computer interaction, Internet of Things (IoT), and wearable technology entered the design vocabulary. The knowledge from these disciplines has also been essential for designers engaging early with the opportunities of additive manufacturing. For example, the software to optimise geometry or introduce a novel lattice structure has, until recently, required advanced engineering or IT knowledge, and even the capacity for writing custom software to do this. Many commercial systems were prohibitively expensive and niche, designed for specific medical or aeronautical applications. Following the trend for 3D printing hardware to become accessible and easier to operate since 2010, currently, it seems that the software and understanding of workflows are finally catching up. Many more options have become available to product designers to focus on exploring the full range of possibilities for 3D printing, rather than being limited to those their tools allowed.

RETHINKING PRODUCT DESIGN AND MANUFACTURING WITH 3D PRINTING

> Technology is humanity's accelerant. Because of technology everything we make is always in the process of becoming. Every kind of thing is becoming something else, while it churns from "might" to "is." All is flux. Nothing is finished. Nothing is done.
>
> [10]

What can be guaranteed is that the next decade will see additive manufacturing become more accessible as a production practice. For technology adoption to be more widespread, its perception for product designers needs to shift from prototyping technology to one for end-use products. While some designers may have engaged with high-end industry 3D printers such as selective laser melting (SLM) or electron beam melting (EBM), it tends to be through service bureaus and external providers where a file is sent to a virtual "black box," and in return a final metal part is received via courier. What occurs within the black box is unknown, and it is quite rare to come across product designers who have actually loaded material into an industrial 3D printer, or pulled parts out of a cake of powder, or filed off support materials from metal 3D prints.

Technicians operating these machines tend to have engineering backgrounds, as do the companies and academic schools that operate them. However, there is no practical reason product designers cannot better engage with industrial 3D printing, just as they have with other industrial manufacturing processes. Hands-on experience is an important part of understanding how to design for the different 3D print processes – removing metal support material by hand is a trial by fire that inspires anyone to optimise their designs to minimise this labour-intensive process. Chapter 1 shared photographs of these processes and detailed these realities, which are often missing from news stories and 3D printer sales material.

While there is an almost limitless supply of content detailing the intricacies of materials and processes for 3D printing, perhaps the greatest challenge for everyone involved with the technology is the pace of change. One day, a new 3D printer is all the rage, being installed in well-known companies around the world; the next day, it is outdated and relegated to the back of a workshop as a backup machine to the new, faster, cheaper, and bigger version. It is with this in mind that this book was written. The three strategies focused on how to transform practice, whether for the individual maker, for a start-up, or for a large multi-national manufacturer. They are technology-inspired, but not technology-dependant, and will help a product designer, or their client, through a journey from least disruptive to completely transformative approach.

The specific guidelines provided through Chapters 1 and 7 support the key considerations for designers working with the most likely 3D printers encountered in manufacturing. They are starting points for the practitioner in working with the technology, and they are far from comprehensive. For comprehensive design guidelines, numerous books [11] and almost limitless journal articles and websites provide insights into every aspect of 3D printing. These range from the ideal geometry and size of powder particles, to spread in powder bed fusion 3D printers, to the difference that printing speed or temperature can play on the ultimate tensile strength of different polymers.

This information cannot possibly be contained within a single book and, unavoidably, dates as quickly as the machines they refer to are constantly updated. They should also only be considered as useful guidelines to begin working with 3D printing where rules can always be broken often to interesting effects. Thus, many guidelines will be based on commercial optimisations to enhance machine throughput, rather than describe the true limitations of a particular technology or material. The prototyping mindset of product designers is perfect for exploring what is possible with the technology and is one of the ways they can add value to current and future understanding of 3D printing. The more product designers engage hands-on with the different 3D printing technologies, the more its capabilities for design can be understood.

3D printing can be another tool in the product designer's toolkit. Knowledge of different forms of printing, what they can and cannot do, and the opportunities and pitfalls they provide is a useful addition to working practice. However, maximising the technology involves stepping back from conventional practice and considering its potential for a different way of operating, both for businesses and, beyond that, for society. The UK Design Council describes its mission as making "life better through design," and the World Design Organisation describes the industrial design profession as "a more optimistic way of looking at the future by reframing problems as opportunities." These position design as inherently future-focussed and proactive. 3D printing as a whole, encompassing the myriad of different processes, presents an opportunity to rethink the relationship between people and products that aligns with the values of the discipline. It is a technology that maximises the use of raw materials, aligned to sustainability strategies, and enables the personalisation of products, aligned to user-centred design. Because it uses digital files, it can help reduce warehousing of inventory, and reduce transport miles.

It is this theme, to rethink traditional ways of manufacture, that links the examples and case studies presented throughout this book. This extends to the most common use of 3D printing within the product design discipline, rapid prototyping, and the opportunity to rapidly iterate, which is the first approach within Strategy 1, "Working with Existing Production." Most product designers will be familiar with 3D printing for this purpose. However, it is important to remember that many businesses, whether manufacturers or otherwise, may not fully grasp the potential of 3D printing to fail fast and fail often. This is the recommended starting point to explore, and it invariably leads a business towards end-use products, even if those products are supporting existing practice such as bridge manufacturing, jigs and fixtures, enhanced tooling, or agile manufacturing, the approaches rounding out Strategy 1 in Chapter 3. Moving through these approaches may happen organically, driven by curious and creative individuals in a company, or may require a product designer to develop examples and business cases for change. In many ways, the specific cases chosen for Strategy 1 are less important than setting up a positive culture around the technology, and the new ways of working that it enables. A top-down push from management to use 3D printing is likely to encounter resistance and raise concerns that "the robots" are going to take people's jobs. Thus, a bottom-up pull from the workers, who are more likely to operate the technology, is far more likely to gain traction. This thinking has informed the logical progression of approaches within the three strategies of this book.

One reason for the slow-uptake of 3D printing beyond the entry-level approaches of Strategy 1 is the tendency to evaluate the viability of commercial 3D printed products on a like-for-like basis. Companies that have successfully interrogated the process into production have exploited the added value that the technology can offer in an evaluation of factors beyond the immediate like-for-like comparison. Part consolidation, for example, may reduce maintenance costs over time and minimise costs associated with manual methods of assembly and sourcing individual components. This is the first recommended approach defined by Strategy 2, "Product Redesign and New Product Design." Light-weighting may reduce the amount of energy needed in any design that involves movement. While more expensive to manufacture than a simpler machined part, over time, this lightweight part could significantly reduce the fuel used in an aircraft with a 30-year service life or improve the speed of an athlete to edge out the competition at the finishing line. Although these are measurable gains, they are rarely factored into current analyses used to justify adoption of 3D printing in end-use practice.

Similarly, a system for customised products may create a more effective product, and one with more emotional buy-in from customers, improving their loyalty and likelihood to maintain and repair a product rather than throw it away.

These benefits are not easy to see in a spreadsheet comparing traditional manufacturing against a proposed 3D printing solution. For 3D printing to become ubiquitous as an option for end-use products, companies need to be informed and prepared for long-term investment and adjustment, optimising their business models as detailed through Strategy 2, leading into Strategy 3, "Digital Business Innovation."

For companies looking to future-proof, additive manufacturing can be a key technology to reduce costs through the digitisation of spare parts inventory, and development of new methods for distribution. These form the first approaches of Strategy 3, extending beyond end-use applications of 3D printing, and setting up new systems for operating in a globally connected society. These are relatively new and uncharted concepts for product manufacturers today. However, lessons can be learned from other domains. For example, the music and film industries have rapidly shifted from selling physical goods (e.g., CDs and CD players, DVDs and DVD players, etc.), to streaming services where customers can watch or listen to what they want, when they want, on any device they want. According to Kevin Kelly, founding executive editor of *Wired* magazine, "In the next 30 years we will continue to take solid things – an automobile, a shoe – and turn them into intangible verbs. Products will become services and processes" [12]. As shown in the Distribution example in the Strategy 3 chapter (Chapter 5), COVID-19 and the community-driven response to personal protective equipment (PPE) shortages around the world highlighted how 3D printers enabled distributed manufacturing of products in close proximity to where and when they are needed. The ongoing consequences of this unexpected use case for 3D printing are not clear. However, it is clear that manufacturers, logistics companies, material suppliers, governments, and consumers finally took notice of 3D printing and the flexibility it provided in uncertain times.

For product designers, flexibility in supply also leads to flexibility in design. Building on the foundations of end-use applications from Strategy 2, Strategy 3 shifts to the more complex approaches of personalisation and scalable systems of supply. These approaches are about digital workflows, with the physical 3D printed outcome simply an output from a point in time, while the system lives on and produces potentially limitless variations over time. For those involved in planning for the future of the product design discipline, this is a field likely to align with the aspirations for the discipline and has been referred to as "meta-design" [13]. Product designers involved in meta-design develop systems that may bring together multiple digital technologies, including 3D scans and/or medical images, combined with automation of workflows and tools described through Strategy 2. These include generative design, topology optimisation, and light-weighting. This bringing together of multiple types of technologies in new and creative ways is known as combinatorial innovation [14]. Where engineers tend to focus on optimising processes for 3D printing, it is this bringing together of technologies, centred around 3D printing, and the development of systems in which people, products, and processes align in sustainable ways, that product design may be defined for the future.

> Today, I realise that design presents a fascinating interplay of technology and psychology, that the designers must understand both. Engineers still tend to believe in logic. They often explain to me in great, logical detail, why their designs are good, powerful, and wonderful. "Why are people having problems?" they wonder. "You are being too logical," I say. "You are designing for people the way you would like them to be, not for the way they really are."
>
> [15]

Arguably, 3D printing is what the profession has been waiting for. In other words, the means to control the manufacture of user-centred products, and to support the exploration of form and function that has fascinated product designers since the inception of the discipline, as consistently illustrated in the case studies of Chapter 6. Additive manufacturing has matured to a point where it is no longer a proposition, it is a reality. Over 30 years it has evolved rapidly, driven by engineering and technical knowledge, bringing together electromechanical hardware and advanced software to produce functional products by adding material in layers. These technical hurdles are now largely overcome. It is time for product design to take up challenge to direct the next 30 years of adoption and applications.

The intent of this book is to provide a pathway to do this. It is a blueprint for cumulatively building skills and knowledge whilst helping to de-risk the adoption of the technology. It provides a basis for building contemporary design conventions and business models for the future.

REFERENCES

1. Gore A. *The Future*. London: WH Allen; 2013.
2. Sparke P. *The Genius of Design*. London: Quadrille Publishing; 2009.
3. Tatum J. The Challenge of Responsible Design. *Design Issues Summer* 2004;20(3): 66–80.
4. Brown T. *Change by Design, Revised and Updated: How Design Thinking Transforms Organisations and Inspires Innovation*. Updated ed. New York: Harper Collins; 2019.
5. Candy S, Potter C, Antonelli P, Burdick A, Dunagan J, Hill D, Jain A, McDowell A, Morton T, Zaidi L. Design and Futures. independently published; 2019.
6. Dunne A, Raby F. *Speculative Everything: Design, Fiction, and Social Dreaming*. Cambridge, MA: Massachusetts Institute of Technology Press; 2013.
7. Lewrick M. *Design Thinking for Business Growth: How to Design and Scale Business Models and Business Ecosystems*. Hoboken, NJ: Wiley; 2022.
8. Ulrich K, Eppinger S. *Product Design and Development*. 6th ed. New York: McGraw-Hill; 2015.
9. Clarke A, Papanek V. *Designer for the Real World*. Cambridge, MA: MIT Press Academic; 2021.

10. Kelly K. *The Inevitable*. New York: Penguin; 2016.

11. Diegel O, Nordin A, Motte D. *A Practical Guide to Design for Additive Manufacturing*. Berlin: Springer; 2019.

12. Kelly K. *The Inevitable*. New York: Penguin; 2016.

13. Giaccardi E. Metadesign as an Emergent Design Culture. *Leonardo* 2005;38(4): 342–349.

14. Nowak P. *Humans 3.0: The Upgrading of the Species*. London: Harper Collins; 2015.

15. Norman D. *The Design of Everyday Things. Revised and Expanded Edition*. Philadelphia, PA: Basic Books; 2013.

Glossary of terms and acronyms

Main 3D printing/additive manufacturing technology terms:

Technology ASTM terminology	Other acronyms or brand names
Material extrusion	FDM: Fused Deposition Modelling FFF: Fused Filament Fabrication BAAM: Big Area Additive Manufacturing
Vat photopolymerisation	SLA: Stereolithography DLP: Digital Light Processing CLIP: Continuous Liquid Interface Production
Material jetting	Polyjet
Binder jetting	
Directed energy deposition	DMD: Direct Metal Deposition LMD: Laser Metal Deposition WAAM: Wire Arc Additive Manufacturing EBAM: Electron Beam Additive Manufacturing LENS: Laser Engineered Net Shaping CLAD: Continuous Laser Assisted Deposition
Sheet lamination	
Powder bed fusion	
Polymer powder bed fusion	SLS: Selective Laser Sintering LS: Laser Sintering MJF: Multi Jet Fusion SHS: Selective heat sintering
Metal powder bed fusion	SLM: Selective Laser Melting DMLS: Direct Metal Laser Sintering DMP: Direct Metal Printing DMLM: Direct Metal Laser Melting EBM: Electron Beam Melting

Other common additive manufacturing related terms, acronyms, and abbreviations

3DP 3D printing

3MF Additive manufacturing file format used to describe colour, textures, materials, and other characteristics of a 3D model. Ongoing development of the file format is led by the 3MF Consortium, which was initiated by Microsoft and other companies in 2015.

AM Additive manufacturing. Process of joining materials to make parts from 3D model data, usually layer upon layer, as opposed to subtractive manufacturing and formative methodologies; historical terms are additive fabrication, additive processes, additive techniques, additive layer manufacturing, layer manufacturing, solid freeform fabrication, and freeform fabrication.

AMF Additive Manufacturing File format for communicating additive manufacturing model data including a description of the 3D surface geometry with native support for colour, materials, lattices, textures, constellations, and metadata.

CAD Computer-aided design; the use of computers for the design of real or virtual objects.

CAE Computer-aided engineering; CAE software offers capabilities for engineering simulation and analysis, such as determining a part's strength or its capacity to transfer heat.

CAM Computer-aided manufacturing; typically refers to systems that use surface data to drive CNC machines, such as digitally driven mills and lathes, to produce parts, moulds, and dies.

CNC Computer numerical control; machines equipped with CNC capabilities include mills, lathes, and flame cutters.

DfAM Design for additive manufacturing.

DwAM Designing with additive manufacturing.

FEA Finite element analysis. Uses computer simulation to predict how forces, fluid, and other properties interact with an object. This can be used to determine if a product will break or meet other performance criteria.

RP Rapid prototyping. Application of additive manufacturing intended for reducing the time needed for producing prototypes. Historically, rapid prototyping (RP) was the first commercially significant application for additive manufacturing and has therefore been commonly used as a general term for this type of technology.

STL Stereolithography file; STL can also stand for Standard Tessellation Language, Standard Triangle Language, or Stereolithography Tessellation Language. This is the file format for 3D model data used by machines to build physical parts. STL is the de facto standard interface for additive manufacturing systems. STL originated from the term stereolithography. The STL format uses triangular facets to approximate the shape of an object, listing the vertices, ordered by the right-hand rule, and unit normals of the triangles, and excludes CAD model attributes.

Support material   Additional sacrificial material which is used to support over-hanging features that could not otherwise be printed, or to act as a heatsink to remove heat from the part being printed. Support material is, in most cases, automatically generated in the correspondent AM machine software.

Topology optimisation   The use of mathematics to optimise the strength-to-weight ratio of a design. The approach minimises the use for a given set of load and constraint conditions.

Notes on organisations, companies, and designers

3D Systems: https://www.3dsystems.com
3DXL: http://www.3dxl.net
ActivArmor: https://activarmor.com
Adrian McCormack: https://www.adrianmccormack.com
Andiamo: https://www.andiamo.io
Anish Kapoor: https://anishkapoor.com
Anouk Wipprecht: http://www.anoukwipprecht.nl
Artek: https://www.artek.fi/en/
Assa Ashuach: https://assastudio.com
ASTM International: https://www.astm.org
Atlas Copco: https://www.atlascopco.com/en-au
Autodesk: https://www.autodesk.com
Barry X Ball: https://www.barryxball.com
Benjamin Dillenburger: https://benjamin-dillenburger.com
Berto Pandolfo: https://profiles.uts.edu.au/Berto.Pandolfo
BMW: https://www.bmw.com/en/innovation/3d-print.html
BOZAR: https://www.bozar.be/en
Carbon 3D: https://www.carbon3d.com
Centre for Sustainable Design, UK: https://cfsd.org.uk
Charlotte Dickson: http://www.charlottedicksondesign.com/stigma-to-silver-linings
Creative Design and Additive Manufacturing Lab, Uni. of Auckland: http://www.
    cdamlab.com
David Trubridge: https://davidtrubridge.com/au
Designboom: https://www.designboom.com
Design Council, UK: https://www.designcouncil.org.uk
Desktop Metal: https://www.desktopmetal.com
DUS Architects: https://houseofdus.com
Eloise Cleary: https://www.uts.edu.au/about/faculty-design-architecture-and-
    building/news/transforming-medical-devices-one-step-time
Emerging Objects: https://emergingobjects.com
Empire Cycles: https://empirecycles.com.au
E-Nable: https://enablingthefuture.org
EOS: https://www.eos.info/en

Fung Kwok Pan: https://sg.linkedin.com/in/fungkwokpan
Futurefactories Studio, Lionel Dean: https://www.futurefactories.com
Forust: https://www.forust.com
GE Additive: https://www.ge.com/additive/
Good Health Design, AUT, New Zealand: https://www.goodhealthdesign.com
Griffith University - Business, Strategy and Innovation: https://www.griffith.edu.
au/griffith-business-school/departments/business-strategy-innovation
Herston Biofabrication Institute: https://metronorth.health.qld.gov.au/herston-biofabrication-institute/
Hewlett Packard (HP): https://www.hp.com/us-en/printers/3d-printers.html
Hyundai: https://www.hyundai.com/worldwide/en/
IDEO: https://cantwait.ideo.com
Iris van Herpen: https://www.irisvanherpen.com
IPCC Climate Change Report: https://www.ipcc.ch/report/ar4/syr/
IMaterialise: https://i.materialise.com/en
James Novak: https://www.edditive.com
Japan Business Federation (Keidanren) Society 5.0: https://www.keidanren.or.jp/en/speech/2019/0101.html
Laika Studios: https://www.laika.com
Legacy Effects: https://www.legacyefx.com
MAD Architects: http://www.i-mad.com
MADE, Victoria University of Wellington: https://www.wgtn.ac.nz/design-innovation/research/industrial-design/made
Materialise: https://www.materialise.com
Maxoniq: https://maxoniq.com
Michael Hansmeyer: https://www.michael-hansmeyer.com
Michelin: https://www.michelin.com.au
Michael Eden: http://www.michael-eden.com
Mimaki: https://mimaki.com
Neri Oxman: https://oxman.com
Nervous System: https://n-e-r-v-o-u-s.com
Nick Ervinck: https://nickervinck.com/en
Objective 3D: https://www.objective3d.com.au
OCED Extended Producer Responsibility: https://www.oecd.org/env/tools-evaluation/extendedproducerresponsibility.htm
ODD Guitars: https://www.oddguitars.com
Orthokids: https://www.orthokids.com.au
Polylab: https://polylab.org
Prosfit: https://prosfit.com
Prusa 3D: https://www.prusa3d.com
Red Dot Museum: https://museum.red-dot.sg
Rehook: https://rehook.bike
Renishaw: https://www.renishaw.com
Reshape 2019: https://youreshape.io/reshape19/

Richard Dupont: https://www.richarddupont.com/projects
Riddell: https://www.riddell.com
Right to Repair Legislation: https://www.europarl.europa.eu/RegData/etudes/BRIE/
    2022/698869/EPRS_BRI(2022)698869_EN.pdf
Ron Arad: http://www.ronarad.co.uk/home/
Ross Lovegrove: http://www.rosslovegrove.com
SCION: https://www.scionresearch.com
Shapeways: https://www.shapeways.com
Spentsys: https://www.spentys.com
Stratasys: https://www.stratasys.com/en
Studio Kite: https://www.studiokite.com
Tom Dixon: https://www.tomdixon.net/en/
UCODO: https://www.dezeen.com/tag/ucodo/
Volvo: https://www.volvocars.com/au
Weta Workshops: https://www.wetaworkshop.com
WIRED: https://www.wired.com
Xkelet: https://n-e-r-v-o-u-s.com

Index